D1431795

TECHNIQUES AND APPLICATIONS
OF THIN LAYER
CHROMATOGRAPHY

Techniques and Applications of Thin Layer Chromatography

Edited by

JOSEPH C. TOUCHSTONE
Department of Obstetrics and Gynecology
School of Medicine
University of Pennsylvania
Philadelphia

JOSEPH SHERMA
Department of Chemistry
Lafayette College
Easton, Pennsylvania

A Wiley-Interscience Publication

JOHN WILEY & SONS
New York • Chichester • Brisbane • Toronto • Singapore

Library of Congress Cataloging in Publication Data

Main entry under title:

Techniques and applications of thin layer chromatography.

    "Third Biennial Symposium on Thin Layer Chromatography
    ...held in Parsippany, New Jersey in December 1982"—Pref.
    "A Wiley-Interscience publication."
    Includes index.
    1. Thin layer chromatography—Congresses.
I. Touchstone, Joseph C. II. Sherma, Joseph. III. Sym-
posium on Thin Layer Chromatography (3rd : 1982 :
Parsippany, N.J.)
QD79.C8T43 1984   543'.08956   84-11924
ISBN 0-471-88017-5

Contributors

Satinder Ahuja
CIBA-GEIBY Corporation
Suffern, New York

Marvin C. Allen
Conoco
Ponca City, Oklahoma

Daniel W. Armstrong
Georgetown University
Washington, D. C.

Rose M. Becker
Indiana University,
    Purdue University
Indianapolis, Indiana

Merlin K. L. Bicking
State University of New
    York
Buffalo, New York

Joel Bitman
U. S. Department of Agri-
    culture
Beltsville, Maryland

Mary S. Bowker
Lafayette College
Easton, Pennsylvania

Udo A. Th. Brinkman
Free University
Amsterdam, Netherlands

Georg de Vries
Free University
Amsterdam, Netherlands

Victoria L. Dimonie
Lehigh University
Bethlehem, Pennsylvania

Mary E. DiPaola
University of Delaware
Newark, Delaware

v

Mohammed S. El-Aasser
Lehigh University
Bethlehem, Pennsylvania

Hansruedi Felix
ANTEC AG
Bennwil, Switzerland

Heinz Filthuth
Berthold
Wildbad, Germany

R. J. H. Gray
University of Delaware
Newark, Delaware

Margit Hamosh
Georgetown University
    Medical School
Washington, D. C.

Paul Hamosh
Georgetown University
    Medical School
Washington, D. C.

Elaine Heilweil
Whatman Inc.
Clifton, New Jersey

Eric K. Johnson
Indiana University,
    Purdue University
Indianapolis, Indiana

Albert J. Kind
Lafayette College
Easton, Pennsylvania

Leslie A. Koska
Lafayette College
Easton, Pennsylvania

Russell K. Lander
Merck, Sharp and Dohme
Research Laboratories
Rahway, New Jersey

Adelaide P. Lee
Lafayette College
Easton, Pennsylvania

Sidney S. Levin
University of Pennsyl-
    vania
Philadelphia, Pennsyl-
    vania

Ted T. Martin
Conoco
Ponca City, Oklahoma

Nitin R. Mehta
Georgetown University
    Medical School
Washington, D. C.

Brian R. Mullin
Uniformed Services
    University of the
    Health Sciences
Bethesda, Maryland

David Nurok
Indiana University,
    Purdue University
Indianapolis, Indiana

Charles M. B. Poore
Uniformed Services
    University of the
    Health Sciences
Bethesda, Maryland

Dexter Rogers
Consultant
Mays Landing, New Jersey

Bruce A. Ruggeri
University of Delaware
Newark, Delaware

Bonnie H. Rupp
Uniformed Services
    University of the
    Health Sciences
Bethesda, Maryland

Ronald M. Scott
Eastern Michigan Univer-
    sity
Ypsilanti, Michigan

Constance S. Seckel
Children's Hospital
Columbus, Ohio

Joseph Sherma
Lafayette College
Easton, Pennsylvania

Barry P. Sleckman
University of Pennsyl-
    vania
Philadelphia, Pennsyl
    vania

Howard R. Sloan
Children's Hospital
Columbus, Ohio

Kimberly A. Snyder
University of Pennsyl-
    vania
Philadelphia, Pennsyl-
    vania

Henry M. Stahr
Iowa State University
Ames, Iowa

Gale Y. Stine
Georgetown University
Washington, D. C.

J. Byron Sudbury
Conoco
Ponca City, Oklahoma

Ronald E. Tecklenburg,
    Jr.
Indiana University,
    Purdue University
Indianapolis, Indiana

Richard C. Tomlins
University of Delaware
Newark, Delaware

Joseph C. Touchstone
University of Pennsyl-
    vania
Philadelphia, Pennsyl-
    vania

Laszlo R. Treiber
Merck, Sharp and Dohme
Research Laboratories
Rahway, New Jersey

John W. Vanderhoff
Lehigh University
Bethlehem, Pennsylvania

Tom R. Watkins
Hunter College, CUNY
New York, New York

David L. Wood
U. S. Department of
    Agriculture
Beltsville, Maryland

Carolyn M. Zelop
Lafayette College
Easton, Pennsylvania

Debra L. Zink
Ohio State University
Columbus, Ohio

Preface

The Third Biennial Symposium on Thin Layer Chromatography was held in Parsippany, New Jersey in December 1982. The proceedings that compose this volume indicate an international participation and envisage the continued growth of thin layer chromatography (TLC). Newer techniques including methods for high-performance and reversed-phase TLC are described. The general advance of the methods is exemplified in the descriptions of the applications and presentation of theoretical aspects of the techniques.

For the first time there is participation from investigators in petroleum laboratories. The methods described should be of interest to other areas of chromatographic analysis since the computerization of the results is presented. Some of the more recent presentations of theoretical aspects of TLC are of interest to all.

Again the meeting could not have succeeded without the help and encouragement of the manufacturers of the requisite supplies. The demonstrations given and presentations by their representatives enhanced the quality of the meeting. We are indebted to all who assisted in the progress of the meeting and to the various participants who generously gave their time in manuscript preparation, leading to the success of the

manuscript preparation, leading to the success of the
volume presented here.

                              Joseph C. Touchstone
                              Joseph Sherma

Philadelphia, Pennsylvania
Easton, Pennsylvania
January 1985

Contents

xi

TECHNIQUES AND APPLICATIONS OF THIN LAYER CHROMATOGRAPHY

CHAPTER 1
Lasers and Thin Layer Chromatography: Present and Future

Merlin K. L. Bicking

INTRODUCTION (THE PRESENT)

The title of this chapter is an obvious variation on the familiar "past, present, and future" theme. However, it is not an unrealistic assessment, since thin layer chromatography (TLC) was first reported as a "new chromatographic technique" in 1941 (1). The laser was invented in the 1960s (2). The fact that it took less than 20 years for two such widely separated fields of endeavor to cross is, then, actually quite remarkable. Since this crossing occurred only a few years ago, it is quite appropriate to conclude that the subject of lasers and TLC really has no "past." On the other hand, it has a very active "present" and certainly a promising "future."

This crossing of fields first occurred in 1975, with the publication by Berman and Zare (3) of an article describing "Laser Fluorescence Analysis of Chromatograms: Sub Nanogram Detection of Aflatoxins." The apparatus described consisted of a pulsed N_2 laser and detection electronics (photomultiplier, boxcar integrator, strip chart recorder, etc.). The TLC plate was manually scanned across the output of the laser and the fluorescence of the aflatoxins monitored. The authors readily admitted that the instrumentation was crude,

but they still realized the potential of the technique
and predicted that an order of magnitude further
improvement in sensitivity could be obtained. As we
shall see later, this prediction was quite true. This,
then, marked the beginning of "the present."

No reports of laser-induced fluorescence appeared
in the literature until very recently (4). As predicted, detection limits for aflatoxins were reduced by
slightly more than an order of magnitude to 10 pg,
allowing detection at the low ppb level in a complex
matrix. The apparatus was constructed so that reproducible measurements could be obtained. It was necessary to solve a variety of problems in order to
achieve the results reported (4), and this chapter will
deal in detail with some of those problems as they
relate to the general use of lasers for quantitation of
species on TLC plates.

It should be noted that there have been other
successful applications of lasers in TLC analysis.
Most unstable among these are techniques involving
photoacoustic spectroscopy (PAS), Raman spectroscopy,
and nonlinear luminescence. These, along with other
applications of lasers, will be discussed later. In
many ways, they mark the beginning of "the future."

We are now prepared to look more closely at current technology involving lasers and TLC--how it got
to where it is, why it got to where it is, and where
it is going in the near future.

ADVANTAGES OF LASER-BASED DETECTION METHODS

It is now common for every article about lasers to
begin with a list of the advantages of a laser over a
conventional light source. At the risk of being
redundant, such a list appears here also. Hopefully,
lasers have lost their image as mysterious black boxes
that emit death rays which blast through buildings.
They are really quite "user-friendly," now that turnkey
systems are available and are common in many laboratories. It is with this fact in mind that once more
(and hopefully for the last time) the unique characteristics of a laser are discussed, but this time as they
apply specifically to TLC.

1. Power

Most lasers will easily produce at least 100 times as many photons as a conventional light source. This is a real advantage in fluorescence spectroscopy, where the intensity (power) of the observed fluorescence is directly proportional to the power of the source (i.e., 100 times as many incident photons means 100 times as many emitted fluorescence photons. Indeed, it is this fact alone that has helped Raman spectroscopy grow. Many other processes that also have low "efficiencies" offer realistic alternatives when a laser is used.

2. Collimated Output

The laser's output needs very little adjustment once it leaves the instrument. It consists of a relatively narrow beam of light that only needs to be focused by one lens, and sometimes focusing is not even necessary. More importantly, the <u>entire</u> output is focused onto the plate. Conventional light sources may be considered a point source, and only a small fraction of the output can be collected and focused. Thus, we make much more efficient use of a laser's output.

3. Monochromatic Output

The spectral line width of the output is typically less than 1 nm, easily comparable to the performance of an inexpensive monochromator as used in commercial TLC scanners. In addition, there is not the extra loss of light caused by passage through the optical components of the monochromator. The development of dye lasers means that these light sources are now also available over an essentially continuous spectral range from 360 to 700 nm.

4. Pulsed Output

Most lasers operate in a pulsed mode, typically producing bursts of photons at rates anywhere from 10 to 1000 Hz. This means that the peak power (that is, the number of photons produced at the maximum of the output) is much higher than conventional sources,

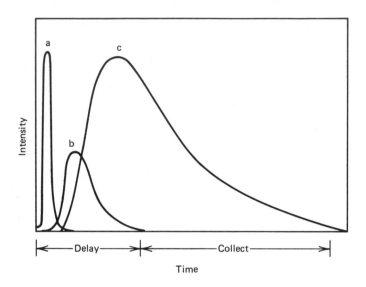

Figure 1. Luminescence intensity vs. time curves for
time-resolved spectroscopy. (a) Laser output. (b)
Background (unwanted) species. (c) Analyte signal.
Collection of light is delayed until unwanted signal
levels have decreased, allowing sampling of the analyte signal with essentially no background.

since a large number of photons are produced over a
time period of about 1 nsec. This characteristic,
coupled with modern electronic technology, allows use
of a technique known as time-resolved spectroscopy.
This is illustrated in Figure 1. Curve a represents the
laser output as a function of time. Curve b represents
the luminescence of some background (unwanted) species;
curve c is the observed luminescence of the analyte of
interest. Note that the two species have different
lifetimes for the luminescence. For this advantageous
situation, it is possible to delay collection of the
light until the background species has decayed to a low
level. Then the analyte luminescence can be collected
during a time "window" when there is very little background interference. This may further lower detection
limits.

5. Cost

This has yet to appear on an "advantage" list, but
now deserves to be here. At least two low-cost tunable
dye lasers are now commercially available (5).   All
they require is benchspace and a tank of nitrogen.
Cost is presently about $10,000, about half the cost of
a good liquid chromatograph. With improvements in sim-
plified detection systems, a laser-based TLC system may
eventually become cost competitive with LC.

DEVELOPMENT OF A LASER-INDUCED FLUORESCENCE
DETECTION SYSTEM

Constructing a laser-based detection system for TLC is
not, unfortunately, a simple matter of putting the two
techniques together and turning on the system. As with
any combination of techniques, trade-offs are necessary, and special arrangements/precautions must be
taken.   Such is the case with lasers and TLC. This
section will deal with some of these problems and their
solutions. Much of the following will deal with the
development of a specific system (4), but many of these
problems and solutions are common when combining the
two techniques, regardless of the exact detection mode.

1. Background

This is often cited as the limiting factor in many
spectroscopic techniques. It may have a variety of
meanings, from chemical to electronic, involving solvent impurities, stray room light, detector or amplifier noise, matrix components, and so on. Unfortunately, all may apply in TLC, along with some additional
problems unique to the TLC experiment. Let us consider
each of these problems in turn.

a. The TLC plate. With the exception of those working
with plates intentionally impregnated with a fluorescing species, most workers have not considered the TLC
plate to be a source of background fluorescence.
Indeed, it has often been mentioned in the literature
that fluoresecence is a particularly sensitive technique because the plate does not fluoresce.

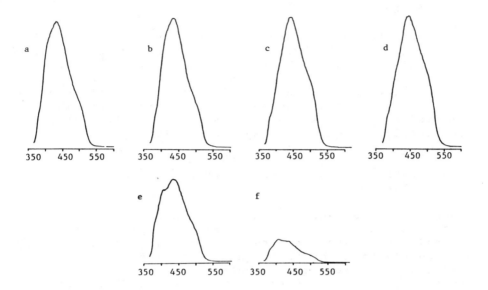

Figure 2. Intensity of fluorescence vs. wavelength (nm)
on a TLC plate.   (a) Aflatoxin B1, (b) B2, (c) G1, (d)
G2, (e) internal standard, (f) plate background.

     Enter the laser.   The most common laser for fluo-
rescence analysis is the nitrogen laser, emitting light
at 337.1 nm.   This "longwave UV" light is ideal for
producing fluorescence in a large number of organic
molecules. With a large number of photons available,
silica gel will also fluoresce, producing sufficient
emission to be seen visually. Optical spectroscopists
have known for some time that glass will fluoresce.
This is one reason why quartz is preferred (in addition
to the fact that glass does not transmit light of this
wavelength very well).
     Figure 2 illustrates the fluorescence spectra of a
TLC plate along with the spectra of four aflatoxins (6).
The plate is seen to exhibit a maximum emission at
about 405 nm, in the blue region. Notice, though, that
this is a broadband emission, and that it completely
overlaps the aflatoxins spectra. It is not possible in
this case to eliminate this source of background. Of

course, any species with fluorescence emission greater
than 550 nm will not be limited by this effect.
Unfortunately, a majority of analytes of interest
fluoresce in the blue region.

    The support for most TLC plates is glass. It must
be realized that this underlying layer will also contribute to the background. It has been observed (6)
that after removing the adsorbent layer, the glass
backing will produce a measurable fluorescence, at a
level roughly 25 percent of the total background (backing and adsorbent). Since the blue fluorescence of
the adsorbent is easily seen from the opposite side of
the plate, it is clear that a significant fraction of
the photons reaches the plate, and may contribute to
the background. It was for this reason that plates
with metal backing were originally used (3). The
future will certainly involve alternative adsorbent
supports.

    The impurities present in the silica (or alumina)
may also pose a problem. The plates used in (4) (Ana-
labs, TLP-099) contained an organic binder. It was
not possible to obtain enough information from the
manufacturer to identify the material, but it was known
to involve a metal-based cross-linking agent. Two
factors are important here. The presence of almost any
organic species under a high light flux is likely to
produce some fluorescence. Second, many metals,
either directly or indirectly, fluoresce. Silica normally contains other trace metals that may also interfere. The solution to this problem lies with the
manufacturers at this point, to rigorously clean the
silica, and to be prudent with the quality and use of
organic binders.

    Simply switching to an inorganic binder such as
$CaSO_4$ may not help. $CaSO_4$ is known to have Mg impurities that may fluoresce more than an organic binder.
In fact, it was observed in the author's laboratory
that plates using a $CaSO_4$ binder were actually inferior
to the organic binder plates, in terms of background
fluorescence. This may or may not be a general phenomenon. (A useful test for the presence of metals is
to spot a solution of 8-hydroxyquinoline (oxine) onto
an adsorbent. Oxine is known to form a yellow fluorescing derivative with many metals, and serves as an
inexpensive probe for trace metals.)

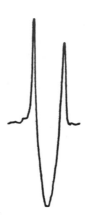

Figure 3. Fluorescence scan across an untreated TLC
plate following application of 5 μl purified methanol.

        The overall background fluorescence level varied
from plate to plate. An intense fluorescent band was
always found migrating with the solvent front (i-amyl
alcohol in this case). The intensity of fluorescence
decreased slowly behind the front and created a changing baseline in the region of the plate containing the
flur aflatoxins. The level and rate of change was
unpredictable and prevented analysis at higher amplifier sensitivities. The purified solvents used exhibited no fluorescence, and tailing was not reduced when
purified solvents were used. Figure 3 indicates that
the impurity was actually present in the adsorbent
material. Following the application of 5 μl of purified methanol to a plate, a fluorescent scan indicated
that the methanol had "eluted" some material. The
"peaks" appeared visually as a circle of radius slightly
smaller than the methanol itself. The lower fluorescence level within the spotted region indicated that
the impurity was indeed in the plate.
        Single or double development with methanol did not
remove the tailing. Continuous overnight development,
however, produced plates with little change in background level. The impurity was seen as a dark yellow
band at the top of the plate. The plates were then
dried in a wormed vacuum desiccator (0.1 mm Hg) for 1

hour. This was preferred to oven drying, which often
produced background "spikes." This is admittedly a
lot of work to prepare the plates, but it is the only
recourse at the moment.

b. Solvents. Of equal concern to plate purity is the
quality of mobile phase used. While the common solvents do not fluoresce, polar, fluorescing impurities
may be trapped, and in effect, concentrated on the
adsorbent. Since almost all fluorescing species are
large, relatively polar molecules, they will undergo a
microlevel solvent demixing phenomenon and be concentrated on the lower portions of the plate. While
normally present at low (η ppm) levels, these impurities will be easily detected by the laser system.
     Reagent grade solvents are not pure enough.
Reagent grade solvents such as chloroform ($CHCl_3$) and
acetone have blue fluorescing impurities. This is
particularly irksome when solutions must be concentrated, thus multiplying fluorescence levels. "HPLC-
Grade" solvents are probably acceptable and should be
used if this is financially possible. However, caution
is necessary since HPLC solvents are specified as
transparent in the UV region (usually below 254 nm),
and the fluorescing impurities are absorbing near 350
nm. Some species may absorb at 350 nm and not at 254
nm, although it is unlikely. An absorption spectrum
would identify any problems immediately.
     For those with budget restrictions, it is better
to buy reagent grade solvents and to purify them. This
can usually be done several liters at a time. Table I
describes purification procedures for several solvents,
modified after reference (7).

2. Optics

     As previously mentioned, the laser's output is
generally suitable for use without much alteration.
Assuming high-quality mirrors are available to direct
the beam as necessary, only one lens is required. Most
workers have used a high-quality quartz lens, which
will give the smallest spot possible (often less than
100 µm). However, developed TLC spots have a finite
width (at least 1 mm), and the use of a small spot is
really wasting most of the sample. While background
levels will increase with the area of contact (width)

TABLE I. PURIFICATION OF SOLVENTS

Solvent	Procedure [a]
Chloroform	Pass through column of Neutral Alumina (Activity Grade 1) or Silica (50 g/1500 ml). Distill.
Acetone	Reflux over $KMnO_4$ 4 hours. Distill.
Acetonitrile	Reflux over $KMnO_4$ 24 hours. Distill, discarding first 10 percent. Pass remaining distillate through Acidic Alumina (activity Grade 1) (100 g/2 1).
Methanol	Dissolve several small chunks Na. Add several grams I_2. Reflux 4 hours. Distill.
Tetrahydrofuran	Reflux 1 hour over CaH_2 or $LiAlH_4$. Distill.
Pyridine	Stir overnight over KOH. Distill over BaO.
i-Amyl alcohol	Dissolve several small chunks Na. Distill.

[a] All distillations should be performed under N_2.

of the beam, the signal for the spot will also increase. Intuitively, it seems reasonable that the laser image should cover the entire spot for best results. This has been verified experimentally (8). Thus, rather than using a spherical lens, a cylindrical quartz lens would be preferred. This allows focusing of the beam to a sharp line, of width on the order of 100 μm.

In order to eliminate stray light and unwanted wavelengths, the laser output should pass through an appropriate filter, (i.e., UV pass only). Similarly,

the collection optics should include a filter which
removes the laser radiation (i.e., visible pass only).
     The collection of fluorescence may be accom-
plished in several ways. A simple two-lens system for
collimation and focusing will suffice. Alternatively,
a slit and detector may be mounted directly over the
plate. However, the lens system as in (4) probably
represents a more stable solution. Further improvements in fiber optics technology may eliminate both of
these approaches.
     The use of a monochromator vs. simple filters
depends on the goals of the analysis, but monochroma-
tors are more flexible in wavelength selection and
should be preferred. The same arguments apply to
photodiodes vs. photomultipliers, with photomultipliers
offering superior sensitivity. The various aspects of
photometric evaluation have been reviewed, and the
paper (9) should be consulted for more details.

     3. Electronics

     At present, overall light levels are not so low
with laser systems as to require special electronic
amplification. Any high-quality electrometer designed
for operation with a photomultiplier should suffice.
However, several unique problems may arise from the use
of a laser excitation source. First, the response time
of the amplifier must be considered. Most lasers
operate at rates of about 20 Hz, so a signal arrives at
the detector every 0.05 sec. This is usually within
the range of most aplifiers, but due to the input
design of many electrometers, the response time will
change with input range settings. The reason is that
each input range has its own time constant (or
response time), a combination of resistor and capacitor,
similar to that used in smoothing many output signals.
The resistor is usually involved in the input scaling,
so each change of sensitivity results in a change in
time constant. Unfortunately, the time constant will
be longer for the more sensitive settings.
     A second problem rests with the nature of the sig-
nal itself. While it may arrive every 50 msec, the
lifetimes of the signal are usually less than 100 nsec,
so all of the signal will arrive in well under 200
nsec (0.20 sec). The amplifier will probably never
"see" this signal because of the input problems, but

will see an "integrated" form of the signal (i.e.,
spread out over a longer time.) Again, if the
response is too slow, these signals will run together,
causing a loss of signal.

Still another problem may result if the response
of the amplifier is fast. The signal arriving at the
detector may have a time-averaged value well within
the range of the equipment (i.e., on a "slow" system
the signal level appears reasonable, e.g. 10^{-8} A).
Remember, though, that the entire signal is arriving in
200 nsec, every 50 msec. This is a concentration factor of greater than 10^5, and the "peak signal" level
can be expected to rise accordingly. This level may be
far above the working level of the detector or amplifier, and "saturation" will result. Again, loss of
signal will result. It is interesting to note that a
few years ago, any detection limits for aflatoxins
below 1 ng would be considered reasonable, with upper
levels in the ug region. In (4), the top of the calibration curve is 5 ng, and at these levels saturation
is observed (in this case due to the detector, not the
amplifier). Clearly, such laser-based methods are
designed for ultratrace analysis, and they are useless
at "high" levels of analyte. An example of the type of
calibration curves obtained is given in Figure 4.

4. Data Acquisition

The amplified signal may be processed in several
ways. If direct copies are desired, a simple RC time
constant can be placed on the amplifier output and the
resulting signal sent to a recorder. The 0.50-sec time
constant used in (4) produced a flat baseline, but it
was necessary to keep the spot scan time at least a
factor of 20 greater than this time (10 sec) to avoid
losing signal. This same smoothed signal was also
directed to the integrator.

It is necessary at this point to stop and further
describe the integrating system. The TLC plate was
mounted on a screw-driven assembly that allowed manual
scanning. Also on the screw drive was a 10-turn heli-
cal potentiometer. The voltage drop across the potentiometer, supplied by a 9-v battery, drove the X-scale
of an X-Y recorder, so that the chart position was
proportional to the TLC plate position. Also mounted
on the screw shaft was a disk containing four equal

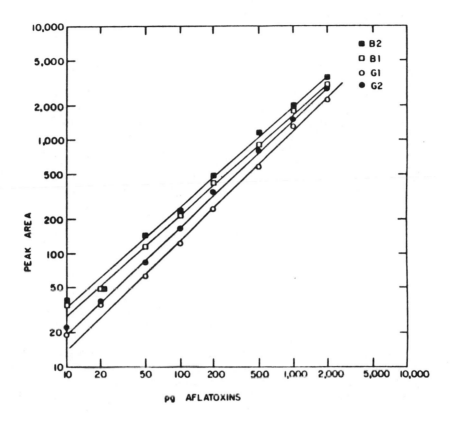

Figure 4. Calibration curve for aflatoxin standards on developed TLC plates.

spaced slots, passing through the cell of an optical interrupter. Every 1/4 turn, the slit passed through the cell, and the interrupter sent a signal to a timer integrated circuit chip (IC 555). The timer IC allowed the voltage/frequency converter to sample for a predetermined time period. The results were dis-played on a counter and summed for each peak.

The unique feature of this system is that it does not operate on a "time base," as most instruments do.

Both recorder position and integrator are activated only by movement of the plate, so that scanning can be stopped and started with no loss of data. This system, then, operates on a "position base," rather than a time base. This is more attuned to the way a chromatographer will envision his plate, and allows considerable flexibility.

A more elegant (and more expensive) modification would involve modern data acquisition equipment and microcomputers such as the popular Apple. A similar arrangement with the optical interrupter should allow the necessary "triggering" to perform data collection at several hundred points along the plate (again, a position, not time base) and perform all the necessary calculations. In either design, it is still necessary to "average" several signal pulses. The limiting factor in data acquisition is the "shot-to-shot" reproducibility of the lasers, which is not much better than a few percent.

5. Other Optical Phenomena

a. Fluorescence Enhancement. It has been observed that spraying or dipping TLC plates in solutions of paraffin oil or glycerol will cause an enhancement of fluorescence (10-12). Flushing with N_2 has also been reported (13) to enhance fluorescence. Figure 5 illustrates the results before and after dipping a plate in 30 percent paraffin oil/hexene. In this case, an enhancement of about 3 was noted, but noise levels also increased so that the overall signal/noise ratio did not change. Apparently, the background luminescence caused by the laser negates the advantage of the oil coating. This has not been reported elsewhere (with conventional sources).

In an effort to further understand this phenomenon, the fluorescence of the aflatoxins was monitored in the presence of a stream of N_2 gas with O_2 added at various levels. The N_2 caused essentially as much enhancement as the paraffin oil. Low levels of O_2 mixed in with the N_2 reduced the fluorescence to original levels. Pure O_2 nearly eliminated all fluorescence. It appears that both oil coating and N_2 enhance fluorescence by a reduction in O_2 quenching. While the techniques do not offer any improvement with

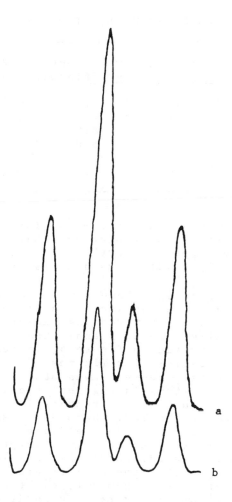

Figure 5. Fluorescence scan of four aflatoxins on a
developed TLC plate: (a) after exposure to 30 percent
paraffin oil/hexane, (b) before exposure.

laser-based systems, they do allow protection of the
plates from atmospheric degradation. Table II shows
the fluorescent lifetimes of the aflatoxins under
various conditions. (This is an example of the usefulness of time-resolved spectroscopy.) No trend is

TABLE II. FLUORESCENT LIFETIMES : (ns)

	O_2	N_2	30 Percent Paraffin Oil	20 Percent Glycerol
Aflatoxin B1	5.3	8.6	5.3	4.4
B2	4.6	7.4	4.5	7.7
G1	6.6	8.8	8.2	6.0
G2	6.1	8.4	8.0	7.2
Plate background	5.7	11.4	6.2	4.6

evident to further elucidate the mechanism of the enhancement, but the values point out that, in this case, time-resolved spectroscopy is of no value, since background and analyte lifetimes are nearly identical.

b. Fluorescence Quenching. Some unusual quenching phenomena have been observed when laser-induced fluorescence with a complex, fluorescing background is monitored (4). While standards exhibited a linear response, as demonstrated in Figure 4, in real samples where the matrix material also fluoresced, only a 50-70 percent "optical" recovery was observed (even though all previous steps in the analytical scheme were shown to give quantitative recovery). Chemical interferences were eliminated. Samples spiked at high levels showed similar recoveries, eliminating any constant mass losses in the method. An absorption type of quenching was indicated, where the background absorbs so much light that the analyte effectively "sees" a lesser intensity. However, dilution of the sample did not increase recovery. In addition, losses were observed using spiked peanut meal, where background levels were barely higher than background. This indicates that there may be a change in a surface characteristic such as the scattering coefficient.

   This particular problem may be caused by both effects, but it points out the fact that lasers may produce both desirable and undesirable results. The quenching problem remains a mystery at this point. Additional work is needed to determine the effect of

high-intensity sources on the scattering properties
of TLC plates. For the moment, an internal standard
is the easiest way to correct for such problems, and
Figure 6 is a classic example of the advantages of
internal standards in analytical chemistry.

EXISTING LASER-BASED TECHNIQUES

As mentioned earlier, laser-induced fluorescence is
not the only laser-based technique reported for TLC.
Several other techniques have been published or are
presently being developed. These are summarized below.

1. Photoacoustic Spectroscopy (PAS)

PAS is a novel spectroscopic technique based on
absorption of light. When a molecule absorbs light,
it may subsequently release the energy as heat (a
nonradiative process). In the PAS experiment, the
sample is held in an enclosed container, and this
released heat will actually create sound waves that
can be detected by a microphone or piezoelectric crys-
tal. The amount of heat produced (and the intensity
of the sound wave) is proportional to the absorbance,
so, in effect an absorption spectrum is obtained when
the source wavelength is varied. The technique is
useful for opaque or light-scattering solids and tur-
bid liquids, which are difficult to analyze normally
(14). The application to TLC is obvious.
Several reports have appeared in the literature
(15-21). In some cases, the work involves physically
removing the analyte and adsorbent layer and trans-
ferring it to a PAS cell (15, 20, 21). These methods
are not really applicable to the present discussion.
Both qualitative (19) and quantitative (15)
information may be obtained. Detection limits rarely
reach the ng level, however. It should be noted that
most PAS experiments utilize a xenon lamp, rather than
a laser. However, the readily available dye lasers
are superior in producing the PAS effect. In this
respect, PAS might be considered as a "future" technique; on the other hand, it is included here because
it is using a pulsed, high-intensity light source.

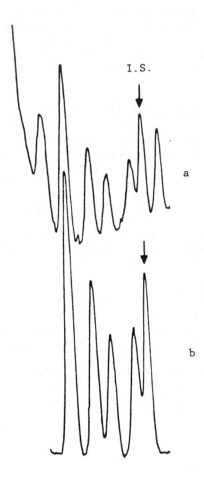

Figure 6. Quenching effects for RDF samples on TLC
plates. (a) Fluorescence scan of an RDF sample
developed on a TLC plate, spiked with aflatoxins and
internal standard. (b) Identical levels of aflatoxin
standards and internal standard developed on the same
plate.

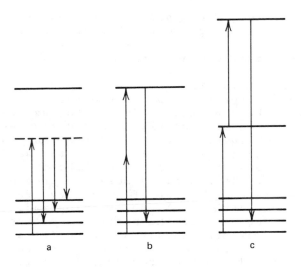

Figure 7. Other techniques for production of luminescence.   (a) Raman spectroscopy--excitation to an
energy level not corresponding to an existing energy
level.   (b) Two photon-excited fluorescence (TPEF)--
simultaneous absorption of two photons.   (c) Sequentially excited fluorescence (SEF)--sequential absorption of two photons to a very high excited state.

   2. Raman Spectroscopy

     Figure 7 illustrates the Raman effect.   Briefly,
a molecule is excited to a "virtual state" (i.e., not
corresponding to any known excited level), and immediately emits a photon of light.   This photon is usually
of a slightly longer wavelength (slightly less energy)
than the source.   This is not a very efficient process,
since most molecules would rather be excited to a discrete energy level.   The large number of photons available is the reason that the Raman effect can even be
noticed (the effect has an efficiency of about 10^{-8}).
Only one report is available of use of the Raman
effect in TLC (22).   With the power available in most
lasers, this technique can produce reasonable signal

levels, but since TLC applications are not common, this technique belongs in the "future" section.

3. Nonlinear Luminescence

The use of lasers on TLC plates can be extended to less common spectroscopic phenomena. Figure 7 also illustrates these effects. Two-photon excited fluorescence (TPEF) and sequentially excited fluorescence (SEF) are techniques recently reported in liquid chromatography detection (23). The TPEF process involves the simultaneous absorption of two photons to reach an excited level. This excited molecule then emits a photon (fluorescence), but of an energy greater than (shorter wavelength than) the exciting radiation. This is the opposite of the normal fluorescence methods. The SEF process involves sequential absorption of photons to a very high excited state, followed by fluorescence. These two processes literally do not follow the "rules" that most of us are taught. It is for this reason that they can be useful, since they offer additional "selectivity" in fluorescence detection, quite unlike any other. These techniques have now been elegantly applied to TLC analysis (24). Since the processes are highly selective, they also tend to be limited in applicability. It remains to be seen how general they are, but detection limits are in the mid pg range for oxadiazole compounds.

4. Photothermal Deflection

Photothermal deflection is a variation of the PAS experiment. Here, one high-power laser is focused on a surface, where a sample spot will absorb the light. As in PAS, the energy is dissipated through heat. In this case, the heat warms up the air over the spot and creates a "thermal lens," an area of gas that is warmer, and has a different refractive index. If a low-power laser is directed through this region, it will be "scattered" when the thermal lens forms. This is the "mirage" effect produced on hot summer days. This is a very recent application (25), and detection limits in the low pg range are reported.

THE FUTURE

If one has enough imagination, there are really no
limits (yet) to the combination of lasers and TLC.
This section, then, is an opportunity to dream a
little.

First, what are the possible limits? Laser technology and optical spectroscopy are not the limiting
factors. We can produce enough photons, and detect a
very few. The surface of the plate itself limits the
techniques. Manufacturers must develop cleaner lots
of silica or alumina, which do not have spectroscopi-
cally active species. In most cases the adsorbent
layer exhibits some response, so its degree of smoothness becomes a factor. Apparently, this is the bigger
technological problem--to prepare a layer that is uniform at the submicrometer level. With the many photons available, a minute change in scattering, caused
by a local irregularity in plate smoothness, transmits
to a noticeable change in signal detected. The solution to this problem is not obvious.

Many common spectroscopic techniques should be
considered. More work, for example, is needed on
applications of Raman spectroscopy. The efficiency of
the process is low, and detection limits are not spec-
tacular. On the other hand, this is probably the only
optical technique with a reasonable response for saturated hydrocarbons. Perhaps fatty acids could be
detected at acceptable levels without derivatization
or charring techniques. A host of simple molecules
not possessing strong absorption bands might be
readily detected, and their structure elucidated, from
Raman spectral data. An additional advantage would be
that Raman spectroscopy is a complementary technique
to IR spectroscopy.

What will be the result of dramatically lowering
the temperature of the plate? How many of the processes will show improved response when the plate is
analyzed at liquid N_2 or He temperatures? Will there
be some advantage to actually <u>heating</u> the plate during
analysis?

Dye lasers are readily available and much work
remains concerning every effect mentioned in this chapter. Without belaboring the point, we need more

information. While these techniques seem to be quite
diverse, they could almost all be performed with a
single system. Dye lasers offer a spectral source
from the ultraviolet through the visible range (comparable to a xenon source). Their pulsed nature allows a
variety of effects, from observing the photoacoustic
effect to time-resolved spectroscopy, where background
and stray light can be eliminated. With suitable mono-
chromators and detectors, one instrument can sample
either direct fluorescence, Raman signals, or nonlinear fluorescence. The development of photodiode
arrays offers the possibility of obtaining a complete
spectrum (for any of the effects) while the plate is
scanned. Microprocessors allow collection, storage,
and evaluation of data. While this "super spectroscopic analyzer" has not been built, there is no
reason to believe it can not be in the near future.

Let us dream a little more. Derivatization may
be extended by in situ photoreactions on the plate.
This would eliminate time-consuming heating procedures.
Consider a developed plate sprayed with the derivatiz-
ing agent. As the laser is scanned across the plate,
it initiates a photochemical reaction with analyte and
spray reagent, forming a derivative instaneously.
Additional photons then cause luminescence by any
desired technique (e.g., fluorescence).

Better yet, why bother to separate the sample
completely (or at all)? If the spectroscopy techniques are specific enough, it may only be necessary
to obtain the most rudimentary separation, followed by
multiple scans, each specific for a certain analyte.
Perhaps then we should spend more time developing
detection systems and less time making plates with
smaller and smaller particles.

This discussion has been admittedly myopic. We
have only scratched the surface with the potential of
the laser, and any spectroscopist can surely name six
more techniques to try. But that is the purpose of
this discussion: to stimulate interest, and to point
to the "future." It would please many workers in the
field to find this discussion obsolete in five years.
If that is the case, it will have been an exciting
five years.

REFERENCES

1. J. G. Kirchner, J. M. Miller, and G. J. Keller,
    Anal. Chem. 23, 420 (1951).
2. B. A. Lengyel, Introduction to Laser Physics,
    Wiley, New York (1966).
3. M. R. Berman and R. N. Zare, Anal. Chem. 47, 1200
    (1975).
4. M. K. L. Bicking, R. N. Kinseley, and H. J. Svec,
    Anal. Chem. 55, 200 (1983).
5. G. M. Hieftje, Am. Lab. 15, 66 (1983).
6. M. K. L. Bicking, Ph.D. Thesis, Iowa State Univer-
    sity, Ames, Iowa (1982).
7. A. J. Gordon and R. A. Ford, The Chemist's Compan-
    ion, Wiley-Interscience, New York, 1972, pp. 431-
    436.
8. H. J. Butler and C. F. Poole, J. High Res. Chroma-
    togr. Chromatogr. Commun. 6, 77 (1983).
9. V. Pollak, in Advances in Chromatography, vol. 17,
    J. C. Giddings (Ed.), Dekker, New York, 1979,
    chapter 1.
10. W. Funk, R. Kerler, E. Boll, and V. Damann, J.
    Chromatogr. 217, 349 (1981).
11. R. Winterstieger and G. Wenninger-Weinzierl,
    Fresenius Z. Anal. Chem. 309, 201 (1981).
12. S. Uchiyama and M. Uchiyama, J. Chromatogr. 53,
    135 (1978).
13. F. DeCroo, G. A. Bens, and P. DeMoerloose, J. High
    Res. Chromatogr. Chromatogr. Commun. 3, 423 (1980).
14. A. Rosencwaig, Anal. Chem. 47, 592A (1975).
15. S. L. Castleden, C. M. Elliot, G. F. Kirkbright,
    and D. E. M. Spillane, Anal. Chem. 51, 2152 (1979).
16. C. M. Ashworth, S. L. Castleden, and G. F. Kirk-
    bright, Anal. Proc. (London) 18, 14 (1981).
17. V. A. Fishman and A. J. Bard, Anal. Chem. 53, 102
    (1981).
18. S. Ikeda, Y. Murakami, and K. Akatsuka, Chem. Lett.
    363 (1981).
19. A. Rosencwaig and S. S. Hall, Anal. Chem. 47, 548
    (1975).
20. L. W. Burggraf and D. E. Leyden, Anal. Chem. 53,
    759 (1981).
21. L. B. Lloyd, R. C. Yeates, and E. M. Eyring, Anal.
    Chem. 54, 549 (1982).
22. J. Von Czarneci and H. W. Hiemesch, Actual. Chim.
    55 (1982).

23. M. J. Sepaniak and E. S. Yeung, Anal. Chem. <u>49</u>, 1554 (1977).
24. P. B. Huff and M. J. Sepaniak, Anal. Chem. (in press).
25. T. I. Chen and M. D. Morris, Anal. Chem. (submitted).

CHAPTER 2

Bonded Stationary Phases for Reversed-Phase Thin Layer Chromatography

Ronald M. Scott

Until a decade ago reversed-phase thin layer chromatography (RP-TLC) was performed by impregnating a support with hydrocarbons or silicone oils. These layers facilitated separations of nonpolar substances and found usage particularly by lipid chemists. However, the technique has some disadvantages. It is difficult to coat the layers reproducibly, there is likely to be incomplete masking of polar groups on the support (which was often silica gel), and the coating tends to dissolve into the mobile phase requiring that mobile phase be saturated with stationary phase before use.

In 1969, reaction of silica gel with chlorosilanes produced chemically bonded nonpolar stationary phases that eliminated many of the problems of the impregnated layers. The mobile phase need not be preconditioned and plates can be used with a reasonable expectation of uniform properties from one analysis to the next. Early bonded phases did not adhere well to plates, resulting in fragile layers, but this difficulty has been overcome (1).

Bonded phases have found extensive application in column HPLC, in fact overwhelming acceptance, but as yet their use in TLC has been more limited. Perhaps

25

the extent of use of the phases in HPLC is the greatest
stimulus at present to the use of bonded RP-TLC plates,
given the increasing use of TLC to model separations
and to select appropriate solvents before putting an
expensive HPLC column at risk with a new separation (2).

For those who have not yet tried RP-TLC, a few
words of description. Samples are dissolved in as
polar a solvent as possible to prevent spreading on the
plate. Water, however, is too polar to wet the plate,
and organic solvents such as methanol or acetone are
most often employed. Before development the zones
should be allowed to dry thoroughly. However, because
of the hydrophobic character of the phase, activation
before use is unnecessary and no problem arises from
exposure of the plate to a humid atmosphere.

The mobile phases employed are usually very simple,
being composed of water and a miscible organic liquid
or modifier, often methanol, acetonitrile, ethanol,
acetone, or dioxane. Very high water concentrations
(above 30-50 percent, depending on the brand of plate)
can damage the layers and thus must be avoided. Dissolving salt in the aqueous layer reduces the damaging
effect and allows higher water concentrations to be
used. The salt has only a small effect on the separations. Some commercial plates employ small uniform
sized particles in preparing the layers, so the advantages of high-performance TLC are available if the samples are applied in very small spots or zones. A variety of techniques can be used to sharpen the bands
front to back after spotting. A very easy one involves
developing the plate for a short distance with a nonpolar solvent. The sample moves at the solvent front
and is concentrated there as a narrow band. The same
effect is obtained by using commercial plates with a
narrow strip of material that is a weak adsorbent, a
preadsorbent or preconcentration area, composed of
keiselguhr or some other weakly adsorbent silica material. Samples can be streaked onto this area with little regard to solvent employed, and with much more aban-
don than is the case on a conventional plate when good
resolution is important. During development, the solutes move rapidly to the edge of the bonded phase and
concentrate there in a narrow zone. When using these
plates, it is also possible to spot much cruder samples
since some preliminary clean up goes on in the preadsorbent zone.

The technique termed multiple development reduces or prevents diffusion of the zones front to back during development of the plate. The plate is developed, dried, and redeveloped, perhaps several times. Each time the solvent rewets the spot it contacts the tailing edge first and moves it toward the leading edge. Using a more polar solvent than would be employed for a single development of the mixture, the components can be moved into the high-resolution zone of the plate stepwise. Other variations in technique can improve resolution on RP-TLC.

Gradient elution has been performed by supplying solvent to a small chamber in which the plate sits using a gradient generating pumping system. The chamber overflows as new solvent enters, providing a continuously changed mobile phase to the plate (3).

Silica gel plates are available commercially which have a strip of bonded nonpolar phase at one end (Figure 1). These allow two-dimensional chromatography to be performed with a reversed-phase separation in the first direction, and conventional phase chromatography in the second. This is an ideal way to obtain the dispersal of solutes which is the goal of the two-dimensional technique, since the mechanism of separation in each direction is different. Scraping enough of the silica gel off to prevent it from dipping into the solvent during the first development prevents it from being deactivated by the aqueous solvent of the first development. When the relatively nonpolar solvent systems used in the second development encounter the solutes on the nonpolar phase, the solutes are rapidly moved to the edge of the silica gel layer, there to be concentrated into sharp bands as had occurred with the preadsorbent plates. The standards for identification may be run in the second direction above the point of maximum development in the first direction.

The reversed-phase technique can be extended to charged molecules by use of ion-pair chromatography. Development of this method is credited to Schill (4), and as with other reversed-phase methods, the greatest application of ion-pair chromatography has been in HPLC. To a typical reversed-phase solvent system a counter ion is added, this being an ionic molecule with substantial nonpolar character. The mechanism of ion-pair separations is visualized as occurring either by the ion-pair forming and associating by the

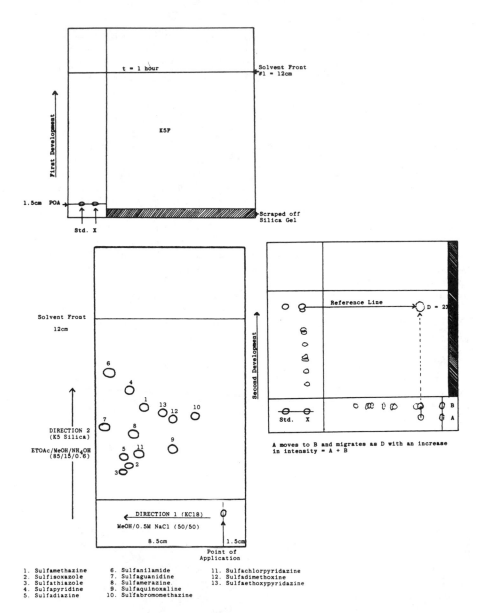

1. Sulfamethazine   6. Sulfanilamide   11. Sulfachlorpyridazine
2. Sulfisoxazole   7. Sulfaguanidine   12. Sulfadimethoxine
3. Sulfathiazole   8. Sulfamerazine   13. Sulfaethoxypyridazine
4. Sulfapyridine   9. Sulfaquinoxaline
5. Sulfadiazine   10. Sulfabromomethazine

Figure 1. Use of bonded (first direction) and conventional phases for two-dimensional analysis. (Courtesy: Whatman Chemical Separation Division, Clifton, N. J.).

TABLE I. EXAMPLES OF COUNTER IONS

Quaternary amines
    N-dodecylpyridinium chloride
    Trimethylpalmitylammonium chloride
    Tetraoctylammonium bromide
    Tetradodecylammonium bromide
Tertiary amines
    Trioctylamines
Aryl and alkyl sulfonates
    Triethanolamine dodecylsulfonate (Na^+ salt too
       insoluble)
    Sodium dodecylbenzenesulfonate
Alkyl sulfates
    Sodium lauryletherosulfate
    Sodium dodecylmydrogensulfate

---

hydrophobic region of the counter ion to the stationary phase, or by the counter ion first associating
with the stationary phase, converting it to the
equivalent of an ion-exchange layer.   See Table I for
the common used counter ions.

$$R^+_{aq} + C^-_{aq} \rightleftharpoons (R^+C^-)_{aq} \rightleftharpoons (R^+C^-)_{org}$$

R^+ = solute
C^- - counter ion

Increased attraction between the ionic solute and the
stationary phase occurs when:

    1. The ability of the counter ion to ion pair is
       greater (functionality of counter ion).
    2. The hydrophobic portion of the counter ion is
       larger.
    3. The concentration of the counter ion is
       increased.
    4. The pH is adjusted to maximize the percentage
       of solute in the ionic form.
    5. When the organic modifier is more polar.
    6. When the concentration of the organic modi-
       fier is reduced.

TABLE II. EFFECT OF COUNTER ION CONCENTRATION[a],[b]

R_f Values

Solute	A	B	C	D	E	F
Aniline	0.70	0.52	0.42	0.30	0.21	0.17
O-toluidine	0.65	0.43	0.36	0.27	0.19	0.15
2,4-dimethyl-aniline	0.56	0.29	0.23	0.13	0.09	0.06
2,6-diamino-toluene	0.85	0.61	0.58	0.41	0.31	0.22

a. Data from Ref. 5.
b. Mobile phase--water : methanol : acetic acid :: 64,3:30:5,7 with ethanol-amine dodecylbenzenesulfonate as counter ion at these concentrations: A, 0%; B, 0.5%; C, 1.0%; D, 2.0%; E, 3.0%; F, 4.0%.

TABLE III. EFFECT OF PREDEVELOPMENT OF PLATE
WITH COUNTER ION ON R_f [a,b]

Number of Developments	R_f		
	A	B	C
1	~1.0	~1.0	~1.0
2	0.41	0.53	0.775
3	0.41	0.52	0.64
4	0.39	0.48	0.62
5	0.39	0.50	0.62

a. From Ref. 7.
b. A, B, C are dyes separated on a Merck RP-8 plate
using methanol : water :: 85:15 with 0.3% tetrahexyl-
ammonium bromide as counter ion.

See Table II for the effect of modifier concentration on R_f values.

Presaturation of the phase with counter ion usually is important to obtain its full effect and to
assure reproducibility. In Table III we see that a
set of dyes moves roughly at the solvent front even in
a solvent-containing counter ion. However, predevelop-
ing the plate once brings the R_f values down sharply,
and after two such treatments of the plate, stable R_f
values result. Figure 2 shows the effect of changes
in methanol concentration on the R_f values.

To devise a mobile phase for ion-pair chromatography in cases where both charged and uncharged species are present as solutes, the system is first optimized in terms of type and concentration of modifier
for the uncharged structures. Under these circum-
stances the ionic material will move approximately to
the solvent front. Then the counter ion is added,
which will have little effect on the other separations,
to a concentration that provides good resolution of
the ionic species.

The mobile phases devised for RP-TLC analysis are
very useful indicators of what works with the corresponding reversed-phase column in an HPLC system (8).

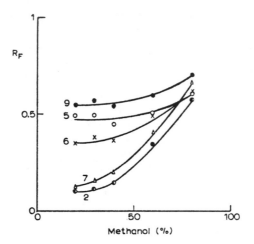

Figure 2. The effect of modifier concentration on solute retention.   From Ref. 6.

The elution order is usually maintained.   In many cases comparison of HPLC capacity factors with $(1/R_f - 1)$ values from TLC results in a linear plot, as does the dependence of log K or Rm on the structure of the solutes.   Two problems arise.   First, RP-TLC cannot deal with solvents as rich in water as can HPLC. Second, there is a basic difference in the operation of the two in that TLC plates are run dry, while HPLC columns equilibrate with solvent before the sample is injected.   It has been reported that the correspondence of RP-TLC and RP-HPLC is even closer for ion-pair separations than for conventional reversed-phase analyses.
      Finally some considerations in the use of Rm values from RP-TLC to predict the partition coefficient of compounds.   Partition coefficients are of interest particularly as predictors of entry by the crossing of membranes into biological systems.   Data from penicillin analyses indicate a somewhat linear relationship between log P and Rm (Figure 3).   Data for phenols (Figure 4) display a trend, but one with considerable scatter.   In the final analysis in some cases, the biological activity of molecules is not badly

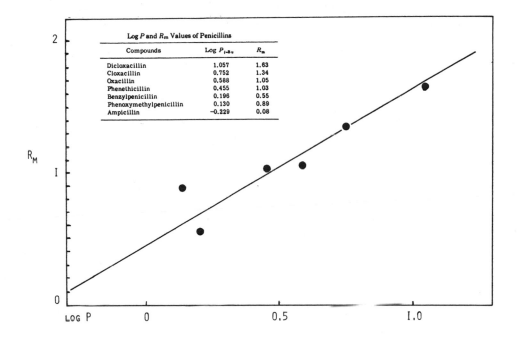

Figure 3. Log vs. R_M values for penicillins. From
Ref. 9.

predicted by Rm values. However, one should not
expect the relationships always to be simple and linear since a good deal more than lipid solubility is
involved in biological effectiveness. In conclusion,
although the data display some correlation, they do
not inspire one to scrap separatory funnels. Data
have largely been gathered using impregnated layers
rather than bonded phases, and interference by
unmasked polar groups of the support has been reported.
Perhaps bonded phases can improve the Log P-Rm correspondence, but important questions as to whether the
mechanism of retention in the nonpolar phase is partition, or is explained as partially or predominantly
an adsorption phenomenon, remain unanswered.

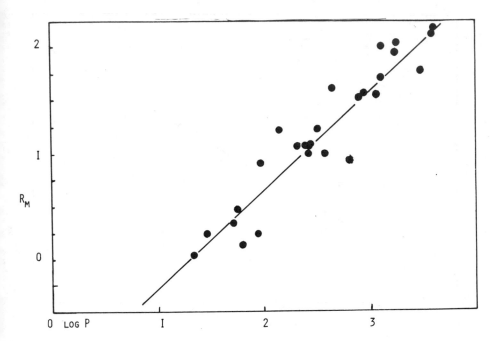

Figure 4.   Log P vs. R_M values for phenols.   From
Ref. 10.

REFERENCES

1. A. M. Siouffi, T. Wawryznowicz, F. Bressolle, and
   G. Guichon, J. Chromatogr. <u>186</u>, 563 (1979).
2. U. A. Th. Brinkman and G. de Vries, J. High Resolut.
   Chromatogr. Chromatogr. Commun. <u>5</u>, 476 (1982).
3. L. C. Sander and L. R. Field, J. Chromatogr. Sci.
   <u>18</u>, 133 (1980).
4. S. Eksborg, P. Lagerstrom, R. Modin, and G. Schill,
   J. Chromatogr. <u>83</u>, 99 (1973).
5. L. Lepri, P. G. Desideri, and D. Heimler, J.
   Chromatogr. <u>155</u>, 119 (1978).
6. L. Lepri, P. G. Desideri, and D. Heimler, J.
   Chromatogr. <u>153</u>, 77 (1978).
7. G. Gonnet and M. Marichy in <u>Recent Developments in
   Chromatography and Electrophoresis</u>, <u>10</u>. <u>Analyti-
   cal Chemistry Symposia Series</u>, vol. 3. A. Frigerio
   and M. McCamish (Eds.), Elsevier, Amsterdam, 1980.

8. S. Roggia and G. G. Gallo, ibid.
9. G. L. Biagi, M. C. Guerra, A. M. Barbaro, and
   M. F. Gamba, J. Med. Chem. 13, 511 (1970).
10. G. L. Biagi, A. M. Barbaro, O. Gandolfi, M. C.
    Guerra, and G. C. Forti, J. Med. Chem. 18, 868
    (1975).

CHAPTER 3

A Systematic Approach to Mobile Phase Design and Optimation for Normal and Reversed-Phase Chromatography

Elaine Heilweil

Traditionally, mobile phase selection has been a trial and error process. Even though solvent theory was being elucidated, few <u>practical</u> benefits of the theoretical advances percolated down into the average laboratory. There was, of course, the elutropic series table to which the practitioner could refer in the process of developing a solvent mixture, but essentially it was still a time-consuming process of elimination until a suitable mobile phase was arrived at.

Selecting the stationary phase is less complicated simply because the selection is more limited. One may have to make a decision to use a normal phase or reversed-phase chromatography plate. Once that decision is made, there are relatively few different types of sorbents available, so that the choice is made simply by being limited. Selecting the <u>mobile</u> phase, however, is more difficult and time-consuming, since the possible combinations available are infinite.

The work of Snyder (1,2) and others enables the chromatographer to approach the problem of mobile phase design in a systematic way, by exploring the selectivities of widely divergent solvents and gathering a vast amount of information in a short time. Snyder classified solvents into eight general groups

TABLE I. SNYDER'S CLASSIFICATION
OF SOLVENT SELECTIVITY

Group	Solvents
I	Aliphatic ethers, tetramethylguanidine, hexamethyl phosphoric acid amide, (trialkyl amines)
II	Aliphatic alcohols
III	Pyridine derivatives, tetrahydrofuran, amides (except formamide), glycol ethers, sulfoxides
IV	Glycols, benzyl alcohol, acetic acid, formamide
V	Methylene chloride, ethylene chloride
VI	(A) Tricresyl phosphate, aliphatic ketones and esters, polyethers, dioxane
	(B) Sulfones, nitriles, propylene carbonate
VII	Aromatic hydrocarbons, halo-substituted aromatic hydrocarbon, nitro-compounds, aromatic ethers
VIII	Fluoroalkanols, M-cresol, water (chloroform)

(Table I), basing their selectivity on their ability to interact as proton donors (x_d), proton acceptors (x_e), or dipoles (x_n). This may be represented graphically as a selectivity triangle (Figure 1), each apex of which represents 100 percent of a specific selectivity characteristic. Thus solvents in group I, for example, would have weak proton donor (strong proton acceptor) characteristics and exhibit weak to moderate dipole interactions with the solutes. Conversely, solvents in group V (methylene chloride, ethylene chloride) will interact as strong dipoles and show weak proton donor and proton acceptor characteristics. The

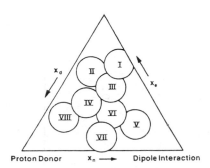 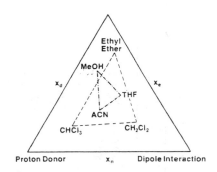

x_e = Proton Acceptor Contribution
x_d = Proton Donor Contribution
x_n = Strong Dipole Contribution

‒‒‒‒‒‒‒ Reversed Phase Chromatography
‒‒‒‒‒‒‒ Normal Phase Chromatography

Figure 1. (a) Snyder's solvent classification triangle. Proton acceptor. (b) Selectivity triangle. Proton acceptor.

contributions of these interaction characteristics for a given solvent add up to unity:

$$x_e + x_d + x_n = 1.0$$

To obtain a change in selectivity in a given separation, the strategy is to choose solvents from opposing apices of the selectivity triangle, since one can expect these solvents to produce the greatest selectivity differences due to their divergent interaction characteristics. Thus for a hypothetical normal phase separation, the solvent trio of ethyl ether (group I), methylene chloride (V), and chloroform (VIII) can be used (Figure 1).

Based on Snyder's theory and subsequent work of Glajch et al. (3) and Youngstrom (4) a seven-solvent scheme can be calculated for a given separation that permits the rapid investigation of possible solute-solvent interactions and the resulting selectivity.

The general procedure is as follows:
1. Choose three solvents from the selectivity triangle which possess selectivity parameters as divergent as possible.
2. Run a binary system by mixing any of the above with hexane (hexane is considered to

TABLE II. SOLVENT STRENGTH DATA: ALUMINA AS ADSORBENT

Solvent	ε^0 [a]	Solvent	ε^0 [a]
Fluoroalkanes	-0.25	Ethylene dichloride	0.44
n-Pentane	0.00	Methyl ethyl ketone	0.51
Isooctane	0.01	1-Nitropropane	0.53
Petroleum ether	0.01	Triethyl amine	0.54
n-Decane	0.04	Acetone	0.56
Cyclohexane	0.04	Dioxane	0.56
Cyclopentane	0.05	Tetrahydrofuran	0.57
1-Pentene	0.08	Ethyl acetate	0.58
Carbon disulfide	0.15	Methyl acetate	0.60
Carbon tetrachloride	0.18	Diethyl amine	0.63
Xylene	0.26	Nitromethane	0.64
i-Propyl ether	0.28	Acetonitrile	0.65
i-Propyl chloride	0.29	Pyridine	0.71
Toluene	0.29	Dimethyl sulfoxide	0.75
n-Propyl chloride	0.30	i, or n-Propanol	0.82
Benzene	0.32	Ethanol	0.88
Ethyl bromide	0.35	Methanol	0.95
Ethyl sulfide	0.38	Ethylene glycol	1.1
Chloroform	0.40		
Methylene chloride	0.42		

a ε^0 = solvent strength

have a zero solvent strength for normal
phase TLC) to obtain an optimum R_f range
(0.2-0.8) and, therefore, an optimum solvent strength P'). This is Solvent 1.
(See Table II for solvent strengths.) For
a mixture of ethyl ether-hexane (50:50)
the solvent strength P' may be calculated
using the formula:

$$P' = F(A) + F(B)$$

where P' = desired solvent strength
      F = volume fraction of component
      A = solvent strength of component A
           from Snyder's table
      B = solvent strength of component B

In the present example, the solvent
strength of Solvent 1 will be

$$P' = 0.5 \times 2.8$$
$$= 1.4$$

because the strength of hexane is zero and
that of ethyl ether is 2.80.
  3. Calculate the remaining two binary systems
     (Solvents 2 and 3) of equivalent solvent
     strength.
  4. Calculate the composition of the remaining
     four solvents of the seven solvents of the
     seven-solvent scheme from a resolution tri-
     angle (Figure 2) in which peaks A, B, and
     C represent the three binary solvents cal-
     culated in Steps 2 and 3.
     Referring to Figure 2 and assuming that the
desired R_f range is achieved with ethyl ether hexane
(50:50) as Solvent 1 with P' = 1.4, calculating compositions as described above will result in the following
seven mobile phase mixtures.
  1. Ether-hexane   (50:50)
  2. $CHCl_3$-hexane   (34:66)
  3. CH_2Cl_2-hexane   (45:55)
  4. Ether-$CHCl_3$-hexane   (25:17:58)
  5. $CHCl_3$-CH_2Cl_2-hexane   (17:23:60)
  6. Ether-CH_2Cl_2-hexane   (25:23:52)
  7. Ether-$CHCl_3$-CH_2Cl_2-hexane (17:11:15:57)

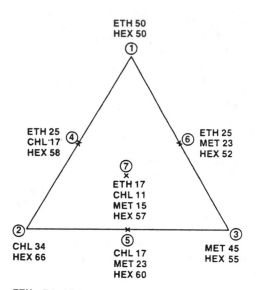

ETH = Ethyl Ether
CHL = Chloroform
MET = Methylene Chloride
HEX = Hexane

Figure 2. Seven-system resolution triangle for ethyl ether, chloroform, methylene chloride (P' = 1.4).

For reversed-phase TLC an identical procedure is followed, using the typical polar organic solvents and water, which in reversed-phase has zero solvent strength, as the diluent.

A good example of the applicability of this procedure to routine mobile phase design in the typical laboratory is the separation of a six-component mixture of sulfonamides. The initial approach was to separate the compounds on a reversed-phase KC-18 plate. Optimum R_f range was obtained by developing an initial chromatogram in methanol-0.5M NaCl (50:50) (Solvent 1) with a solvent strength P'.

$$P' = 0.5 \times 5.1 = 2.55$$

Using methanol (group II), acetonitrile (ACN) (group

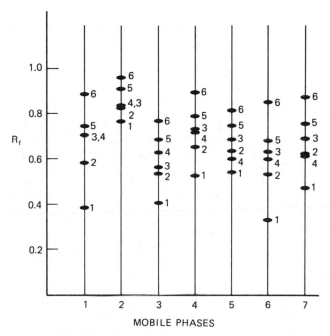

Figure 3. Selectivity chart seven-solvent scheme on KC18 for sulfonamides.

VI), and tetrahydrofuran (THF) (group III) as the solvent trio to be investigated, mixtures of equivalent solvent strength P' were calculated for acetonitrile and THF. The remaining four solvent compositions were calculated using a resolution triangle. The resulting seven mobile phases were:

1. Methanol-0.5M NaCl	50:50
2. THF-H$_2$O *	64:36
3. ACN-0.5M NaCl	44:56
4. MeOH-THF-0.5M NaCl	25:32:43
5. ACN-THF-0.5M NaCl	22:32:46
6. MeOH-ACN-0.5M NaCl	25:22:53
7. MeOH-RHF-ACN-0.5M NaCl	17:21:15:47

* THF does not yield a clear solution when mixed with 0.5 M NaCl at this ratio.

Upon examination of the resulting seven chromatograms, we observe (see selectivity chart in Figure 3)
that although no single mobile phase completely
resolved the six components, the effort yields a
wealth of chromatographic information. It can be
seen that all components may be resolved by using two
different systems. It can also be seen that selectivity changes with acetonitrile-containing systems, so
that sulfapyridine (3) and sulfisoxazole (4), which
are not resolved with systems 1 or 2, are well separated in Solvent 3.

Notice should be taken of the fact that, although
the calculated solvent strength P' for all seven solvents is the same, the R_f values obtained for a given
compound in each system vary. Snyder's solvent theory
does not account for solvent/stationary phase interaction, or solute/stationary phase interactions.
Therefore, the equations that are used here are only
approximations, since an assumption is made that no
such interactions are taking place.

When a desired separation cannot be obtained on a
particular stationary phase, one can attempt the separation on another, with a different separation mechanism.

Calculating a seven-solvent scheme for a normal
phase system for the sulfonamide mixture (Figure 4),
seven new chromatograms are obtained. Here greater
selectivity changes can be seen, with numerous cross-
overs of bands (resolution chart, Figure 5). Again
no single mobile phase separates all six components,
but we can see great selectivity differences between
Solvent 7 and Solvent 6. Solvent 7 does not resolve
sulfadimethoxine (1) from sulfaguanidine (6) or sulfisoxazole (4) from sulfathiazole (5). Both pairs are
separated in System 6 at the expense of other compounds, which System 1 does separate. Somewhere
between these two solvent compositions, one should
arrive at a combination that resolves all six compounds.

To investigate this possibility, mobile phases 1
and 6 are mixed at various ratios, and chromatograms
are obtained with each mixture. By plotting mobile
phase composition versus R_f for each compound (Figure
6), an optimum composition is arrived at that can
separate all components.

SOLVENT OPTIMIZATION

SEVEN — SOLVENT SCHEME FOR SULFONAMIDES

TLC BASED ON SNYDER'S SOLVENT CLASSIFICATION

1)   ETHYLACETATE / MEOH / NH₄OH   (90 / 10 / 1)

2)   CHLOROFORM / MEOH / NH₄OH   (63 / 37 / 1)

3)   ETHYL ETHER / MEOH / NH₄OH   (27 / 73 / 1)

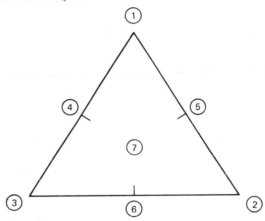

4)   ETHYL ETHER / ETHYLACETATE / MEOH / NH₄OH   (14 / 45 / 41 / 1)

5)   ETHYLACETATE / MEOH / CHCL₃ / NH₄OH   (45 / 23 / 32 / 1)

6)   ETHYL ETHER / CHCL₃ / MEOH / NH₄OH   (14 / 31 / 55 / 1)

7)   ETHYLACETATE / ETHYL ETHER / CHCL₃ / MEOH / NH₄OH   (30 / 9 / 21 / 40 / 1)

Figure 4. Solvent optimization. Seven−solvent scheme for sulfonamides TLC based on Snyder's solvent classification.

Following this procedure separation of all six sulfonamides was obtained on a preadsorbent LK6 plate in a mobile phase consisting of 33 percent of mobile phase 6 in mobile phase 1.

This method is useful in the development of mobile phases for HPLC using TLC, and was presented as such by Youngstrom et al. (4). Compounds do not have to be completely resolved on a TLC plate in order to be separated by HPLC, due to the column's greater efficiency.

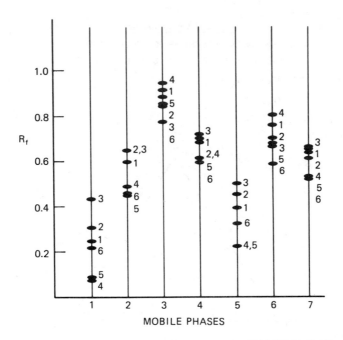

Figure 5. Selectivity chart seven-solvent scheme on K6 for sulfonamides.

        Because of the following relationship between R_f and k',

$$k' = \frac{1 - R_f}{R_f}$$

for a separation to be directly transferrable from TLC to HPLC, R_f values should be in the range of 0.2-0.3 (Figure 7).
        In a laboratory where solvent design work is done routinely, it is possible to accumulate a "library" of seven-solvent schemes according to their solvent strengths, to be reused as the need arises, thereby saving time needed for calculation. Thus, for example, the seven-solvent scheme used for the sulfonamide

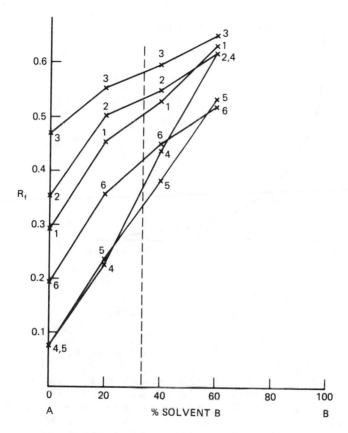

SOLVENT A: ETHYLACETATE/METHANOL/AMMONIA (90/10/1)
       B: ETHER/CHLOROFORM/METHANOL/AMMONIA (14/31/55/1)

Figure 6. Optimizing mobile phase composition by
extrapolation sulfonamides on LK6.

separation was applicable to chromatography of chlorpromazine for a purity determination (Figure 3), with
Solvent 3 giving optimum resolution of chlorpromazine
and impurities.
    In summary, the solvent design method presented
here is:

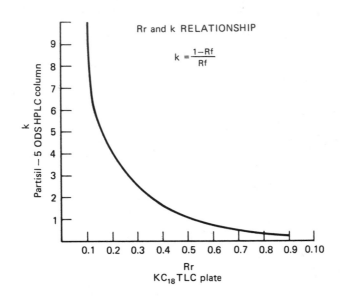

Figure 7. R_f and k relationship.

1. A practical routine way to optimize
   separations.
2. Allows the technician to accumulate a
   library of solvent schemes, which can be
   used and reused, and evaluated in a minimum amount of time.
3. It is a time efficient method of acquiring
   a wealth of chromatographic data.
4. Although based on a complex theory, the
   procedure is easy to implement for routine
   use.
5. The procedure is applicable to mobile
   phase design for HPLC.

    Sherma and Charvat (5) have used this solvent
selection approach to design a mobile phase for the
separation of organophosphorus pesticides on a C_{18}
layer.

REFERENCES

1. L. R. Snyder, J. Chromatogr. Sci. 16, 223 (1978).
2. L. R. Snyder, J. W. Dolan, and J. R. Gant, J. Chromatogr. 165, 3 (1979).
3. J. L. Glajch, J. J. Kirkland, K. M. Squire, and J. M. Minor, J. Chromatogr. 199, 57 (1980).
4. R. Youngstrom, TLC Section, Eastern Analytical Symposium, Nov. 1981.
5. J. Sherma and S. Charvat, Amer. Lab. 15(2), 138 (1983).

CHAPTER 4
Rotating-Disk
Thin Layer Chromatography

Debra L. Zink

INTRODUCTION

Rotating-disk thin-layer chromatography (RDTLC) uti-
lizes centrifugal force to move a solute through a
thin layer of sorbent. The advantages of the addition
of centrifugal force to classical separation systems
are an increase in solvent flow rate with a resultant
decrease in retention time, the ability to use sorbents of finer particle size, thereby increasing system efficiency, and a design that is inexpensive and
simple to use. The addition of centrifugal force to
a separation technique is not new. In 1935, Hickman
(1) introduced centrifugal distillation, which was followed by the development of centrifugal paper, column,
and thin layer techniques. This work was reviewed by
Deyl, Rosmus, and Paulicek (2).
    The renewed interest in RDTLC is due to a major
change in instrumentation. The instrument used in
this study was the CLC-5 manufactured by Hitachi LTD
(Figure 1). The major improvement was the placing of
the TLC sorbent between two plates. The upper disk is
made of stainless steel and tempered glass. Four
screws are spaced evenly on the outer edge of this
disk, and a holding reservoir for solvents is located
in the exact center of the disk. The lower disk is

51

Figure 1. The Hitachi Model CLC-5 rotating-disk thin layer chromatograph. From right to left: the instrument electronics console (containing motor, detector, and fraction collector controls), the chromatograph (the motor is housed in the base), the detector module, and the fraction collector.

stainless steel, the bottom of which has a stem that is used to hold the entire disk on a rotor. A porous spacer fits securely into a groove around the perimeter of the top side of the bottom disk. On the outer edge, there are four holes into which the screws from the upper disk fit, thereby holding the plates together.

This sandwich design not only allows easy packing and uniformity of sorbent layer thickness, but more importantly prevents evaporation from the disk. Control of evaporation allows for increased rotational speeds and better reproducibility of retention times.

The disk is packed by pouring a slurry of sorbent into the center of the disk while it is spinning. This procedure results in the uniform distribution of the sorbent throughout the disk. It has been found

Figure 2. Illustration of the separation of dyes by
RDTLC. Photograph of a plate midway during elution of
a mixture of tetraphenylcyclopentadienone (red), Ama-
plast yellow AGB, fatty orange, indophenol (blue), and
Amaplast orange LFP. Solvent: dichloromethane.
Adsorbent: Baker grade Silica 7.

that plates can be packed reproducibly. Any given
plate has a lifetime of greater than 48 hr if it is
kept wet with solvent.
    Samples are introduced to the sorbent through the
central hole while the disk is rotating at a slow
speed. This results in a slug type of injection that
is useful in preparative applications since a large
volume of solute can be introduced directly to the
sorbent.
    The solutes are moved through the plate by
increasing the rotational speed. Each solute moves
radially and is separated in the form of a concentric

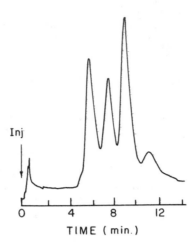

Figure 3. Resultant chromatographic trace at 1000 rpm
rotational speed.

band as it moves through the plate (see Figure 2).
When it reaches the end of the plate, it is washed down
a collector basin to the detector and then into the
fraction collector. Figure 3 shows the densitometric
trace of this separation.

SYSTEM EFFICIENCY

The number of theoretical plates (N) achieved with
Baker grade Silica 7 increased with rotational speed.
The maximum number of theoretical plates at 1000 rpm
(the upper limit of the CLC-5) was 550 with a 2-mm
spacer and 1100 with a 5-mm spacer. In none of the
systems investigated was the form of the van Deemter
curve (3) usually associated with liquid chromatography observed. This could indicate that higher rotational speeds (greater than 1000 rpm) are needed to
achieve the optimum efficiency. However, because of
excess dead volume between the plate and the detector,
this observation may only be a consequence of an
increase in the efficiency in washing the solutes from
the collector basin to the detector.

SYSTEM CAPACITY

The capacity of the system increases with the size of the spacer. With a 2-mm spacer there was a large reduction in the efficiency of the system when 2 to 3 mg of a dye mixture was introduced. Several hundred mg of this same mixture was successfully separated with a 5-mm spacer.

APPLICATIONS

As illustrated previously, the system has been used to separate dye mixtures. A chloroform extract of spinach was fractionated with the same system. Large quantities of crude oil were fractionated by a step gradient elution. Fingerprint patterns were observed for several petroleum products, and further studies by gas chromatography and modern liquid chromatography were made easier by this initial fractionation.

CONCLUSIONS

RDTLC has been overshadowed by recent advances in modern liquid chromatography. However, if one considers the ease of operation, relatively low cost of the instrument/packing material, and the large preparative capacity of the system, RDTLC can be seen to have considerable utility in the fractionation and cleanup of complex samples. This instrument should prove to be very useful in many areas of chemistry, including petroleum studies, natural product characterization, and synthetic chemistry.

REFERENCES

1. K. C. D. Hickman, Ind. Eng. Chem. 29, 968 (1937).
2. Z. Deyl, J. Rosmus, and M. Paulicek, Chromatogr. Rev. 6, 19 (1964).
3. J. J. Kirkland (Ed.), Modern Practice of Liquid Chromatography, Wiley-Interscience, New York, 1971, chapter 1, pp. 11-38.

CHAPTER 5

Time-Optimized Thin Layer Chromatography in a Chamber with Fixed Plate Lengths

Ronald E. Tecklenburg, Jr.,
Rose M. Becker
Eric K. Johnson
David Nurok

INTRODUCTION

Thin layer chromatography (TLC) has many advantages over high-performance liquid chromatography (HPLC) for the separation of simple mixtures. These include the fact that multiple samples can be run simultaneously, that solvent properties do not interfere with solute detection, and that the volume of solvent used per sample is generally 1 to 2 orders of magnitude less than in HPLC provided that a suitable development chamber is used.

A limitation of TLC is that solvent path length is limited to about 20 cm with conventional TLC plates and to about 10 cm with high-performance plates. These path lengths are often not long enough for difficult separations. Such separations can often be accomplished by continuous development whereby solvent is allowed to evaporate off the end of the TLC plate.

Reprinted, with permission, from Anal. Chem. <u>55</u>, 2199 (1983).

It has been recommended that binary solvents compris-
ing a weak solvent such as hexane and a stronger solvent such as acetone be used for these separations (1).
Apparatus for such developments--the Regis Short Bed/
Continuous Development (SB/CD) chamber--has been available for several years and has been used for a variety
of separations.   It has recently been shown that continuous developments with binary solvents can be time-
optimized (2).   The suitability of using the Regis
SB/CD chamber for time-optimized separations is evaluated below.

   EXPERIMENTAL SECTION

The Regis SB/CD chamber was used for chromatography.
The chamber is rectangular with four ridges along the
floor, all parallel to its long axis.   This allows a
TLC plate to be inserted silica gel coated side up in
five positions with the base of the plate resting
against a ridge (or in position 5 on the wall) and the
upper end of the plate propped against the opposite
wall.   The line of contact between the TLC plate and
the glass cover of the chamber defines the line where
solvent starts to evaporate.   It is essential to allow
about 1.5 cm of plate to extend beyond this contact
line as this is the area from which solvent evaporates.
The rate of evaporation can be roughly optimized by
placing the chamber in a hood and raising the front to
an appropriate level.   In our laboratory an opening
height of 18 in. was found satisfactory.   With 25 ml
of solvent in the chamber, the path lengths for positions 1 through 5, from solvent source to the initial
point of solvent evaporation, are 2.35 cm, 3.85 cm,
5.40 cm, 7.00 cm, and 8.30 cm.
   Analtech, Inc. (Newark, Delaware) silica gel
plates, catalogue no. 47011, were cut into appropriate
sections 10 cm wide before use.   Plates were stored in
the laboratory atmosphere overnight before use.
   The humidity reported ahead refers to the ambient
laboratory humidity.
   For the experiment in which plates were stored at
a controlled humidity the plates were spotted and then
placed in a desiccator over 39 percent sulfuric acid
for 15 hours.   This concentration of sulfuric acid
maintains the relative humidity at 60 percent.

Temperature in the laboratory was in the range of 23-25° C.

The steroids used in this study were obtained from Sigma Chemical Co. (St. Louis, Missouri). The solvents used were obtained from Aldrich Chemical Co. (Milwaukee, Wisconsin).

RESULTS AND DISCUSSION

The method of optimization has been fully described elsewhere (2). Familiarity with the original publication is assumed in the following discussion. The method is dependent on using a binary solvent such that the following relationship exists between solute capacity factor, k, and the mole fraction, X_3, of the stronger component of the binary:

$$\ln k = a \ln X_3 + b \qquad\qquad (1)$$

where a and b are constants characteristic of each solute in the binary. It is necessary to determine the constants a and b for each solute by performing TLC at several mole fractions of the binary. It is also necessary to experimentally construct a plot of the solvent velocity constant, κ vs. X_3. Once this has been done, it is possible to calculate at any mole fraction, X_3, the time by continuous development, t_1, necessary to yield a specific center-to-center spot separation, S_D. This is

$$t_1 = (2lL - l^2)/\kappa \qquad\qquad (2)$$

where t_1 is the development time in seconds, l is the length of plate in millimeters, used for continuous development, L is the length of plate in millimeters, required for the same separation by conventional development, and κ is the solvent velocity constant. The length L is a function of the difference in R_f values, ΔR_f, of the pair of compounds to be separated, and of S_D, the center-to-center spot separation required, whereas the length l depends on the product $R_f L$, where R_f refers to the spot of highest R_f value. In the case of a mixture containing more than two components (not considered here) L would be determined for that pair of compounds most difficult to separate.

The value of t_1 varies with solvent composition.   The
minimum of a plot to t_1 vs. X_3 yields the solvent composition at which the most rapid separation occurs.
This value is referred to as $(t_1)_{min}$.
    Time optimization by continuous development
requires a development chamber where any predicted
plate length, l, can be used.   In contrast, the SB/CD
chamber has five fixed positions so that for many separations the correct plate length can only be approximated.
    The analysis time in the SB/CD chamber can be
optimized in the following way.   At a given mole fraction, the analysis time, t_1*, using a plate position
of fixed length, l*, is

$$t_1* = (2 l*L - (l*)^2)/\kappa \qquad (3)$$

The values of L and κ are determined in exactly the
same way as when using eq (2).   The value of l* corresponds to the length of one of the fixed plate positions in the SB/CD chamber.   The minimum of the plot
of t_1* vs. X_3 will determine the solvent composition
that will yield the shortest analysis time for a given
plate position in the SB/CD chamber.   It should be
noted that this minimum analysis time will usually be
at, or close to, the intersection of this plot with
the t_1 vs. X_3 plot obtained by using eq (2) provided
that the same values of slope and intercept are used
for both poles.
    At the mole fraction corresponding to the inter-
section of the two curves the values of l and l* are
identical.   At lower mole fractions the value of
l* > l and the compound of highest R_f will not reach
the top of the plate although the required value of S_D
will still be obtained.   At higher mole fractions the
value of l > l* and the required value of S_D cannot be
obtained as the fixed path length is too short for adequate separation.   Thus the permissible solvent concentration range in the SB/CD chamber is such that l* $\geq$ l.
    We have found that one of the three shortest posi-
tions generally yields the lowest analysis time for a
single pair of compounds.   This includes difficult
separations where a plate length of up to 39 cm (see
Table II) would be required for the same separation by
conventional TLC.   The two longer positions may however be of value for the separation of more complex

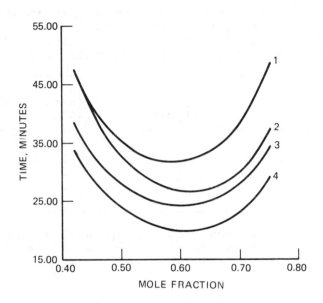

Figure 1. Plots of computed analysis time vs. mole
fraction for a spot separation of 5 mm for hydrocorti-
sone/corticosterone on silica gel with ethyl acetate/
1,1,2-trichlorotrifluoroethane. Slopes and intercepts
are for a humidity of 60 percent. Each plot is computed by use of the slope and intercept corresponding
to one of the four fixed positions in the SB/CD
chamber.

mixtures. The slopes and intercepts (constants $\underline{a}$ and
$\underline{b}$ in eq 1) vary with plate position. Thus even if the
plate length were adjustable, different $\underline{t}_1$ vs. $\underline{X}_3$
curves would be obtained in each position. This is
illustrated in Figure 1, which shows the variation of
$\underline{t}_1$ with the solvent composition for the separation of
corticosterone/hydrocortisone on silica gel by using
ethyl acetate/1,1,2-trichlorotrifluoroethane as solvent and by using the slopes and intercepts found for
positions 1 through 4. It should be emphasized that
the plate length required for $\underline{t}_1$ as defined in eq (2)
varies with the solvent composition, even though this
variation can be slight. Thus the minimum of the low-
est curve (i.e., $(\underline{t}_1)_{min}$, the shortest analysis time)

TABLE I. ANALYSIS TIME AND PLATE LENGTH IN SB/CD
CHAMBER COMPARED TO FULLY OPTIMIZED TLC FOR A 5-mm
SEPARATION OF HYDROCORTISONE/CORTICOSTERONE DEVELOPED
IN ETHYL ACETATE/1,1,2-TRICHLOROTRIFLUOROETHANE

SB/CD Position	l^*, mm	l at $(t_1)_{min}$, mm	$(t_1)_{min}$, min	$(t_1^*)_{min}$, min
1	23.5	39.6[a]	31.5[a]	108.8
2	38.5	34.8	26.5	27.0
3	54.0	31.3	19.7	30.9
4	70.0	33.3	24.1	46.9

[a] Values of $(t_1)_{min}$ and l at $(t_1)_{min}$ are calculated
from the corresponding value of slopes and intercepts
obtained for each of the four fixed positions in the
SB/CD chamber.

in Figure 1 will in general not be attainable unless
there is fortuitous agreement between the variable
plate length l corresponding to the minimum of the
curve and the fixed plate length l^*.

Furthermore, the curve that has the lowest minimum is not always an accurate indicator of which position of the SB/CD chamber will yield the shortest
analysis time for a given separation. This is because
the values of l and l^* can be very different for a
given slope and intercept. This is illustrated in
Table I.

The conditions for time optimized TLC appear to
be dependent on humidity. Figure 1 was constructed at
a relative humidity of 60 percent. Similar figures
were constructed at relative humidities of 21 percent
and 23 percent. The sets of curves obtained at either
of the latter two humidities are virtually identical
with each other but are significantly different to
that obtained at 60 percent relative humidity. The
slope and intercept of position 3 yielded the lowest
analysis time (19.7 min) of the four positions at 60
percent relative humidity whereas at either 21 percent
or 23 percent relative humidity, the slope and intercept of position 1 yielded the lowest analysis time

(24.8 min) of the positions considered. Thus optimized continuous development in the SB/CD chamber
using silica gel plates should be performed at a controlled humidity. Preliminary experiments indicate
that small fluctuations in humidity can be compensated
for by storing the TLC plates under controlled humidity until immediately before use. Plates stored at a
relative humidity of 60 percent gave substantially the
same results at an ambient relative humidity of 54 percent as at an ambient relative humidity of 60 percent.
It is recommended that if this procedure is followed
the composition of the mobile phase be restricted,
such that $\underline{l}^* \geq \underline{l} + 5$ mm. This allows for the possibility of small changes in $R_{\underline{f}}$ due to small fluctuations
in temperature or humidity. Small changes in $R_{\underline{f}}$ do
not have a significant effect on $S_{\underline{D}}$.

The analysis times quoted in the above paragraph
are of course hypothetical as they would only be
obtained if the plate length could be optimized. As
noted earlier this length can be very different from
the length of the corresponding fixed position for
which slope and intercept were measured.

It is clear from the above discussion that the
SB/CD chamber has two drawbacks, namely, that the
slope and intercept values (a and b in eq 1) differ
according to which position is used and that the user
has five fixed plate lengths to choose among. It is
common to find that the optimum length is only approximated by one of these fixed lengths. The latter can
be overcome by adjusting the solvent level. However,
preliminary experiments indicate that this leads to
further changes in slope and intercept values.

In spite of the above shortcomings the SB/CD
chamber is of use in continuous development, provided
conditions are optimized. This is illustrated by the
series of analyses reported in Table II. The agreement between predicted and experimental analysis time
is within 10 percent for all the separations. Moreover, for three of the five analyses reported, development time in the SB/CD chamber is significantly less
than that required for conventional development.

The advantage of working under fully optimized
conditions where both plate length and mole fraction
are completely variable, as compared to partially optimized conditions as in the SB/CD chamber, is illustrated by the two right-hand columns of Table II.

TABLE II. COMPARISON OF TLC IN

PREDICTED FOR CONVENTIONAL

		Solute
	Hydrocortisone/ corticosterone	Hydrocortisone/ corticosterone
$\underline{S}_D$ predicted, mm	5.0	10.0
$\underline{S}_D$ found, mm	10.0	
Solvent used [a]	1	1
Conventional TLC		
$\underline{L}$, cm	20.19	39.89
$(\underline{t}_L)_{min}$, min	69.7	272.0
Optimized TLC		
$\underline{l}$, cm	3.51	5.83
$(\underline{t}_L)_{min}$, min	24.7	84.9
SB/CD chamber		
$\underline{l}*$, cm	3.85	3.85
$(\underline{t}_1*)_{min}$, min	25.8	173.2
$\dfrac{(\underline{t}_1*)_{min}}{(\underline{t}_L)_{min}} \times 100$	37.0	63.7
$\dfrac{(\underline{t}_1)_{min}}{(\underline{t}_L)_{min}} \times 100$	35.4	31.2

THE SB/CD CHAMBER WITH RESULTS

TLC AND OPTIMIZED TLC

Pair		
Progesterone/ pregnenolone	Estrone/ progesterone	Estrone/ progesterone
4.4	4.0	5.0
4.3	4.4	5.4
2	2	2
29.84	29.18	36.22
147.4	140.9	217.2
3.81	4.15	4.94
35.2	37.2	55.2
3.85	3.85	3.85
39.5	52.3	188.4
26.8	37.1	86.7
23.9	26.4	25.4

These express $(t_1)_{min}$ and $(t_1^*)_{min}$ as percentages of $(t_L)_{min}$, the minimum analysis time by conventional development. The latter corresponds to the minimum of the t_L vs. X_3 curve where $t_L = L^2/\kappa$. For the separations listed, the time required under fully optimized conditions is between 24 and 35 percent of that required for conventional development. A lesser reduction in analysis time is found for the SB/CD chamber where the time required is between 27 and 87 percent of that required for conventional development. As noted earlier in this report the reduction in time in the SB/CD chamber is significant in three of the five cases considered. The largest reduction in analysis time is for progesterone/pregnenolone where analysis time is reduced from 147 min for conventional development to 40 min by optimized development in the SB/CD chamber.

Two of the pairs separated are reported at different values of S_D. In both instances, the higher S_D requires a substantially longer time in the SB/CD chamber. Shorter times could possibly be obtained in positions other than position two. This would of course entail measurement of slope and intercept values in each of the new positions.

It should be noted that for the separation of progesterone/pregnenolone there is a 10 percent difference between the fully optimized time and the time in the SB/CD chamber even though there is an insignificant difference between 1 and 1^*. This is because there is only a very slight dependence of plate length on mole fraction in this region for this particular separation.

The pairs listed in Table I are difficult to separate by use of the solvents listed. It should be noted that for four of the five separations, conventional development requires a TLC plate in the range between 26 and 40 cm in length whereas position 2 of the SB/CD chamber (as shown in Table I) requires an overall plate length of about 5 cm. Thus apart from considerations of shorter analysis time, the SB/CD chamber allows difficult separations to be performed that would not normally be obtainable by conventional TLC.

It should be noted that for several of the solute pairs in Tables I and II, $1 > 1^*$. This inequality is permissible because $(t_1^*)_{min}$ and $(t_1)_{min}$ occur at different mole fractions.

Registry No. Hydrocortisone, 50-23-7; corticosterone, 50-22-6; progesterone, 57-83-0; pregnenolone, 145-13-1; estrone, 53-16-7.

REFERENCES

1. J. A. Perry, J. Chromatogr. <u>165</u>, 117 (1979).
2. D. Nurok, R. M. Becker, and K. A. Sassic, Anal. Chem. <u>54</u>, 1955 (1982).

CHAPTER 6

Spectral Analysis
of Thin Layer
Chromatography Bands

Henry M. Stahr

INTRODUCTION

A survey of the current stage of sensitivity for separation techniques for toxicology by thin layer chromatography shows at least three major problems for confirmation of TLC bands by spectroscopy. They are the
resolution, the visualization, and the removal or isolation of bands. Spectral confirmation requires a
means of introduction into the spectrometer with a minimum of decomposition and loss and, also, a minimum
interference from the chromatographic media. Table I
shows the present levels of sensitivity of various
methods used for quantitation of toxic substances.

EQUIPMENT AND REAGENTS

Experimental

Spectroscopy equipment available at the Iowa
State Veterinary Diagnostic Laboratory chemistry facil-
ity includes a Finnigan GC/MS, Varian 219 UV/visible
spectrometer, Beckman IR4 infrared spectrometer, Per-
kin Elmer infrared spectrometer, Aminco Bowman spectro-
flurophotometer, and other spectrophotometric equipment with dedicated use. On the Iowa State University

TABLE I. SENSITIVITY OF SPECTRAL READOUTS
FOR SOME TOXIC SUBSTANCES 1982

Spectroscopy	Sensitivity, μg (Molecular Weight 200–400 AMU nonchlorinated)
Mass spectrometry	10^{-1}(trichothecenes)
Ultraviolet spectrometry	10^{-1}(aflatoxin)
Fluorescence spectrometry	10^{-3}(aflatoxin)
Nuclear magnetic resonance (FT)	100
Infrared (FT) spectrometry	10^{-1}

campus, a Bruker FT-NMR and an IBM FT-IR are available
on a service basis from the Chemistry Instrument Service. A Kontes densitometer is available for quantitative analysis of TLC bands. The following plates were
used for thin layer chromatography: normal phase silica gel plates, with and without fluorescent indicator,
by Analtech, Merck, and Whatman; preparatory plates
from Merck (glass and aluminum backed silica gel
plates), with and without fluorescent indicator; Whatman reversed phase, with and without indicator; and
high-performance plates from Merck and Whatman. Solvents used were distilled in glass (Burdick and Jackson) or Nanograde (Mallinckrodt).

PROCEDURES AND RESULTS

Mass Spectrometry

Analysis by thermal desorption is done after
extracts are recovered from normal-phase TLC plates by
methanol-water (70:30). This eluting solvent works
well for aflatoxins, coumarin rodenticides, and most
of the organic compounds that would move at the solvent front with it as a mobile phase for TLC. Special

considerations are the complete removal of silica gel
(normal or reverse phase) adsorbent and scrupulously
clean surfaces for all operations. Cleaning procedures for thin layer chromatography media and equipment are described below. Levels of sensitivity vary
with the compound and efficiency of the mass spectrometer and the thermal introduction system.

Millipore submicron Teflon® filters are used to
remove silica gel particles from samples, or alternatively, centrifugation (50,000 rpm) or settling overnight. Often the latter is best. Extracts may be
also eluted through a Sep Pak C_{18} column with a very
polar solvent such as methanol with a removal or particles without loss of analyte. Analytes such as
Roundup®, which are significantly soluble only in
water, should be removed by water.

Ramaley (2) has been able to introduce gel with
analyte directly into the mass spectrometer source and
obtain good spectra. Our laboratory is anticipating
more use of this technique with the advent of the 4500
source added to our Finnigan 4000 GC/MS instrument.
Conventional electron impact ionization in the presence of silica gel gives decomposition of nearly any
analyte. With clean chromatography and the removal of
residual particles before thermal introduction of the
analyte into the mass spectrometer source, comparable
spectra are obtained from pure materials and bands
from TLC. Sensitivity ranges from nanogram quantities
for chlorinated hydrocarbons and alfatoxin. Figure 1
shows the mass spectrum of Vomitoxin. The nuclear magnetic resonance spectrum of T-2 toxin is shown in
Figure 2.

DENSITOMETRY

Direct readout of optical density, fluorescence quenching, or fluorescence is possible with the Kontes densitometer. Its filtered light source allows rudimentary
isolation of spectral bands. This can make possible
selenium analysis at the nanogram level and lower with
TLC and densitometry, below the level of fluorometry.
Fluorometry we will deal with next, because no solvent
is necessary to allow the measurement. Kratos (Schoeffel), Shimadzu, and Farrand all make scanners that
allow monochromator-controlled excitation and emission
measurements directly from TLC plates.

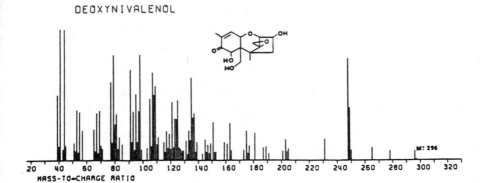

Figure 1. Mass spectrum of vomitoxin.

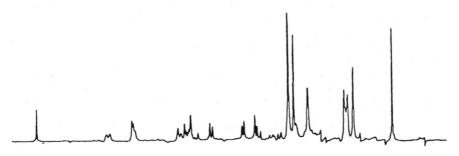

Figure 2. Nuclear magnetic resonance analysis at the milligram level.

FLUOROMETRY

The principal limitation to sensitivity in fluorometry is the sample size required for the fluorometer. Bands that contain nanogram amounts are used for afla- toxin analysis, with milliter size cells. Eighty to 100 times less sample can be used in a 80-μl micro- cell. These cells allow the excitation and emission spectra to be recorded to help provide specificity and allow the measurement of spectra of substances even in the presence of other compounds. Packed cells that use an inert filler have been used to limit volume and

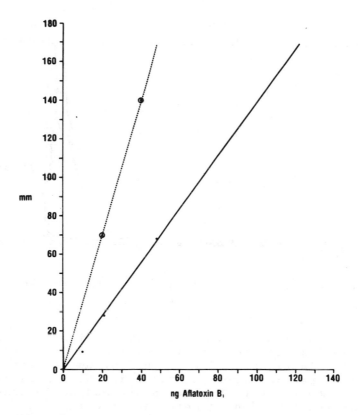

Figure 3. Fluorescence analysis at the nanogram level.

maximize fluorescent light output.   Figure 3 shows the
calibration curve for direct scans of aflatoxins separated on thin layers.

        ULTRAVIOLET SPECTROMETRY

Direct UV spectrometry on thin layer plates has been
reported by Fenimore (3) at a previous biennial symposium.   He reported excellent sensitivity and was
able to analyze very low levels of drugs and biologically active compounds.   The instrument that allows
such work is one of a kind, as far as can be determined (Zeiss Optical $50,000+ cost).   It is now

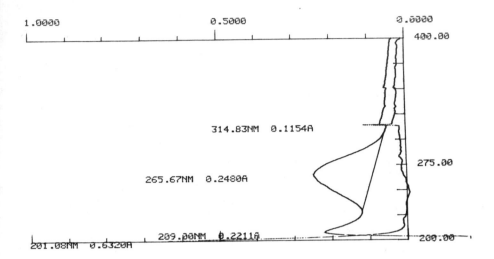

Figure 4. Ultraviolet spectra at the submicrogram
level.

obsolete and no longer manufactured. Normal UV analysis is done like fluorometry in cells. Fluorometry is
more applicable because of the need for purity with
ultraviolet spectrometry. Extracts in which aflatoxin
can be seen by the eye on a TLC plate and concentrations measured by a densitometer and fluorometer give
no discernible spectra by UV spectrometry because of
interferences. Microcells and modern spectrometry can
push the sensitivity to the nanogram/milliliter level
but here the selectivity obtained in fluorometry by
the dual monochromator (excitation, emission) is missing, so bands must be very well resolved. Reversed-
phase chromatography often resolves similar compounds
that will not separate by normal phase TLC, thereby
allowing identification of the analyte of interest in
the matrix. Figure 4 shows the ultraviolet spectrum
of an unknown at the nanogram level after separation
by TLC.

CLEANUP OF EQUIPMENT

Stahr (4) reported in a previous symposium that cleanup of chromatographic equipment was critical and could

be accompanied by solvent extraction, acid bath, and/
or ignition of glassware to annealing temperatures.
After the combustible compounds are removed by ignition or acid bath procedure, it is necessary to rinse
the glassware in Millipore® grade water, or the equivalent, until clean. Adding a double deionized water
rinse allows the use of an automatic dishwasher to
rinse pipets and other glassware. The glassware is
either rinsed by hand or by the same technique used
for the pipets. Chromatographic plates to be used for
separations that will result in spectroscopic confirmation are washed in methanol and then dried and developed in the solvent used for the analytical separation.
     The use of the thermal wire desorption source
with the Finnigan GC/MS instrument or field desorption
with the Varian, Kratos, or Nuclide mass spectrometers
allow picogram sensitivity for even hard to analyze
compounds such as aflatoxins. Nonvolatile antibiotics
and proteins have been analyzed by these systems.

DISCUSSION

The sensitivity of mass spectrometry with MS/MS (mas
spectrometry/mas spectrometry) (three quadrapoles) or
two double-focusing magnetic sector stages plus a
mixer has been enormously increased. The price of
these units is also three times more than conventional
mass spectrometers. Use of these present "state of
the art" units obviates most of the precautions that
have been described here. For quite a while the use
of the more conventional techniques will require these
precautions to be taken. Deconvolution programs for
ultraviolet spectroscopy are now state of the art and
will help with interfering materials. The present sensitivity of 200 ng for Fourier transform intrared spectrometry makes it an extremely useful technique for
confirmation of functional groups in analyses. FT/NMR
so far is only two or three times more sensitive than
that level presented in the talk 12 years ago by
Widmark. It still is not a trace technique for low-
level chromatography.

CONCLUSION

Spectrometry techniques may be used to qualitatively confirm the bands from TLC plates and to quantitatively measure the mass contained in the fraction.

REFERENCES

1. Gunnar Widmark, "Possible Limits of Ultramicro Analysis," Advances in Chemistry Series 104, ACS, Washington, D. C., 1971.
2. L. Ramaley, "Direct Mass Spectral Analysis of TLC Bands." Presented at 28th annual ASMS meeting, New York, May 1980.
3. D. Fenimore and C. Davis, in TLC, Environmental and Clinical Applications, J. C. Touchstone (Ed.), Wiley, New York, 1980, p. 114.
4. H. M. Stahr, D. Lerdal, and W. Hyde, J. Liq. Chromatogr. 4, 1097 (1981).
5. H. M. Stahr, W. Hyde, and L. Sigler, letter to editor, Anal. Chem. (in press), (1982).

CHAPTER 7
Quantitativeness of Scanning Densitometry

Dexter Rogers

The analytical procedure for trace organic substances
in natural matrices involves five steps: (a) sampling
and sample preservation, (b) sample preparation and
cleanup, (c) resolution of a mixture of analytes, usually by some form of chromatography, (d) detection and
quantitation of the resolved analytes, and (e) data
reduction, record keeping, and quality control. Thin
layer chromatography (TLC) can play roles in three of
these steps. The first and foremost function of TLC
is resolving a mixture of compounds. Once resolved,
these compounds can be eluted for quantitation and
further study. The second and no less significant
function of TLC is providing a stable, neutral medium
to preserve the spatial separation of these compounds.
This arrangement is useful for in situ quantitation by
scanning densitometry and, often overlooked, for performing microchemical studies. The third function of
TLC is providing some degree of cleanup, in addition
to resolution. This function can be provided by two-
dimensional TLC and by use of biphasic silica gel
plates prepared with preadsorbent or reversed-phase
zones.

Development chromatography, that is TLC, offers
the analyst more than what elution chromatography
(high-performance liquid chromatography, HPLC) can,
for either analytical or preparative purposes. It

offers additional advantages over HPLC, because TLC is
rapid and convenient, effective and economical, and
able to employ more varied and sensitive means of
detection.

The purpose of this paper is to review in situ
quantitation by scanning densitometry and, in particular, to answer the question "if spectrophotometry is
universally recognized to be quantitative for a transparent medium, then why is not scanning densitometry
universally recognized to be quantitative for a translucent, light-scattering medium?"

Quantitative alludes to the correspondence
between the amount of chromophore present and the magnitude of the signal emanating from the detector. If
the relationship between chromophore and signal is linear and passes through the origin, quantitativeness is
established. Quantitative also alludes to the efficiency of absorption of incident light by the chromophore and the relation of incident light by the chromophore and the relative absence of complicating factors,
such as light-scattering. In simple photometry, light
rays are treated as if they were straight and parallel.
Light scattering would counter that assumption. If
the efficiency of absorption of incident light
approaches 100 percent and any complicating factors are
shown to be not significant, quantitativeness is also
established. Increased efficiency would contribute to
enhanced sensitivity and selectivity for TLC. Both
forms of quantitativeness must be established.

Light scattering is due to microinhomogeneities
in material density within the optical path (1). It
is influenced by particle size and wavelength of the
incident light. For small molecules, the dominant
form of scattering is Rayleigh scattering, which is
wavelength dependent and isotropic. Rayleigh scattering would be equal in the forward and backward direc-
tions. For large molecules and colloids, the dominant
form of scattering is particulate scattering, which is
relatively wavelength independent and anisotropic.
Particulate scattering would be greater in the forward
than in the backward direction. Because of its particulate nature, TLC sorbents will scatter more light
in the forward direction, although the relative mass
of the sorbent makes light scattering appreciable in
any direction.

The evolution of TLC has depended on reliable instrumentation as well as on the theory, practice, and quality supplies. The scanning densitometer represents a quantum leap forward, because it allows for in situ quantitation in various modes of operation, such as absorbance with either transmitted or reflected light, fluorescence, and fluorescence quenching (2). Their calibration curves are linear and pass through the origin, and are sensitive and reproducible. Since we are not always comfortable when working with light-scattering media, we are seeking to learn (a) how the TLC sorbent affects scanning densitometry, and (b) how efficiently are the measurements made by scanning densitometry.

For this study a Kontes densitometer was employed. It was operated in the transmission mode with illumination from a 4-watt fluorescent lamp, which emitted maximum energy at 366 nm and was positioned below the TLC plate holder. The fiber optic/cadmium sulfide detection assembly was positioned above the TLC plate holder directly against a 2-mm quartz space placed in immediate contact with the densitometer standard. The plate holder was moved electromechanically, while the lamp and detector remained in fixed positions. The densitometer was calibrated with an Eastman Kodak step-wedge (Figure 1), which established its linear capability up to at least 1.20 absorbance units.

A densitometer standard was prepared by a photographic process. It consisted of a sequence of 3.0-mm diameter spots, which had graded opacities corresponding to 10, 25, 40, 60, 80, and 100 arbitrary units. It could be scanned without any dimensional distortion by the densitometer. The data indicated a linear relationship between opacity and signal, which passed through the origin. Clearly, scanning densitometry is capable of obeying Beer's law for a transparent medium.

The densitometer standard was rescanned with an undeveloped TLC plate positioned before or after the standard and facing toward or away from the standard. A Whatman LK6 silica gel plate served as the light-scattering medium. A quartz spacer was placed against the exposed surface of the TLC plate. A typical set of results is shown in Figure 2. The black dots at the base of each peak were drawn in to represent the original location and size of the test spots. The

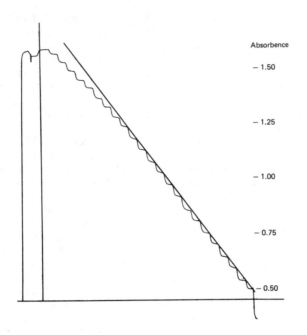

Figure 1. Calibration of a Kontes densitometer with an Eastman Kodak step-wedge.

light-scattering medium broadened the width and low-ered the height of the peaks. The relative opacities of the densitometer standards are plotted against the observed peak heights (Figure 3a) and peak areas (Fig-ure 3b). The results of this experiment are summa-eized in Table I. The peak areas were normalized to opacities corresponding to 100 arbitrary units. These results lead to the following conclusions:

        1. The light-scattering medium, the TLC plate, caused dispersion of light with the result that, when the plate was placed before or after the densitometer standard, the peak widths were broadened from 3.0 to 4.0 mm and 5.0 mm, respec-tively. The peak heights were lowered correspondingly.

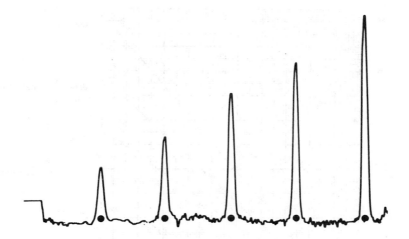

Figure 2. Typical scan with a TLC plate positioned after the densitometer standard.

2. Minor effects were noted when the plate was positioned facing toward or away from the densitometer standard. These effects might have been due to the character of the exposed sorbent surface.
3. The efficiency of scanning was largely retained when the TLC plate was positioned <u>before</u> the densitometer standard. In that position, the TLC plate was affecting the light source much like a neutral density filter, although changing the direction of some light rays.
4. The efficiency of scanning was reduced when the TLC plate was positioned <u>after</u> the densitometer standard. The apparent opacities were reduced by 25 and 50 percent, because light was probably being backscattered into the shadows cast by the opaque spots.

What is being measured in this experiment is a transparent light path extended by the thickness of a light-scattering medium, the TLC plate. In actual practice, the chromophore is distributed uniformly

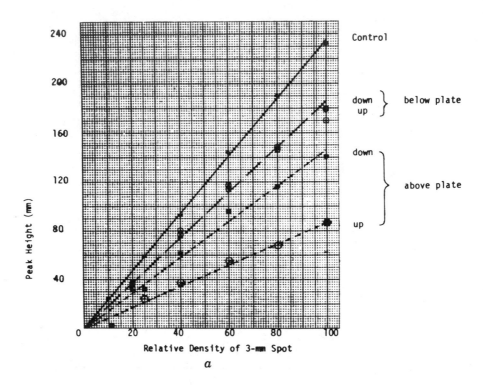

Figure 3(a). Correlation of relative opacity to peak height.

throughout the thickness of the TLC sorbent. Consequently, the light-scattering effect should be only
one-half as great as that observed in this experiment,
because the average light-scattering path is only one-
half of what it was in this experiment. The efficiency of scanning densitometry, therefore, may not be
50 and 75 percent, but actually 75 and 88 percent.
    The distribution of chromophore within the TLC
plate is undoubtedly uniform throughout the thickness
of the plate, because the sample is applied in solution, which penetrates the sorbent or preadsorbent
layer before evaporating, and because scanning densitometry yields the same result whether scanned from
one side or the other. The assumption of uniformity

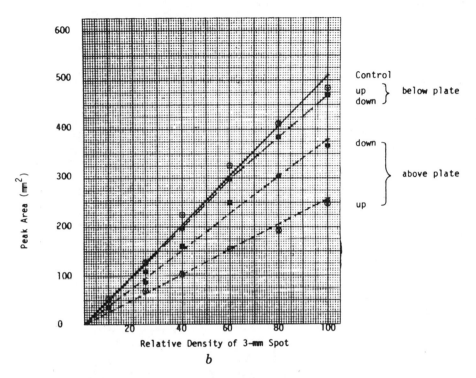

Figure 3(b). Correlation of relative opacity to peak area.

of distribution of material within the spot of application is one of the underlying requirements for TLC (3).
     The practice of measuring absorbance by reflected
light, instead of transmitted light, would not avoid
the light-scattering effect, unless the chromophore
was coating the front surface of the TLC plate.
Reducing the thickness of the sorbent layer would not
reduce the light-scattering effect, because the light-
absorbing path would also be reduced. The net effect
is the same. We know that scanning densitometry of a
light-scattering medium obeys Beer's law: the calibration curves are linear and pass through the origin.
We now suggest that the efficiency of scanning densitometry is reassuringly high, high enough to reduce

TABLE I. SUMMARY OF RESULTS SHOWING THE EFFECT OF
POSITION OF THE TLC PLATE
RELATIVE TO THE DENSITOMETER STANDARD [a]

Test Configuration	Normalized Peak Area	Observed Peak Width
Control: test plate alone	512 ± 8 (1.00)	3.0 (1.00)
Light-scattering medium placed:		
Below the test plate		
Face down	475 ± 28 (0.93)	4.0 (1.33)
Face up	522 ± 29 (1.02)	4.0 (1.33)
Above the test plate		
Face down	382 ± 27 (0.75)	5.0 (1.67)
Face up	255 ± 11 (0.50)	5.0 (1.67)

[a] Figures in parentheses refer to relative values with respect to the control, the test plate alone.

lingering doubts about the validity of applying Beer's law to a light-scattering medium.   It would be inter-esting to compare extinction coefficients determined for both transparent and translucent media.

In these days of computer assistance, the ques-tion arises whether scanning densitometer could be extended to higher ranges of analysis, into the non-linear range of response, by incorporating the Kubelka-Munk theory into Beer's law.   However, the real ques-tion is not whether it can be done, but whether it should be done.   The principal function of TLC is to resolve mixtures.   To facilitate resolution, sample application is kept minimal, and minimal applications will fall in the linear range of analysis.   Thinking small contributes to the effectiveness and efficiency of TLC.

REFERENCES

1. P. C. Hiemenz, Principles of Colloid and Surface Chemistry, Dekker, New York, 1977.
2. General Catalogue TG-50a, Kontes, Vineland, N. J. 08360.
3. J. C. Touchstone and M. F. Dobbins, Practice of Thin Layer Chromatography, Wiley, New York, 1978.
4. L. R. Treiber, B. Oertengren, R. Lindsten, and T. Oertengren, J. Chromatogr. 73, 151 (1972).

CHAPTER 8

Thin Layer Chromatography on Chemicallly Bonded Phases —A Comparison of Precoated Plates

Udo A. Th. Brinkman
Georg de Vries

INTRODUCTION

In modern thin layer chromatography (TLC), silica is
the stationary phase material preferred by a large
majority of all workers, and TLC on chemically bonded
phases occupies a rather modest position. This is in
marked contrast with the situation in column liquid
chromatography (LC), where a considerable proportion
of all work is done using chemically bonded phases,
especially nonpolar C18- and C8-modified silicas.
Still, in recent years the widespread use of these
apolar chemically bonded stationary phases in what is
often called reversed-phase column liquid chromatography (RPLC) has promoted interest in their utilisa-
tion in TLC, and a recent review (1) on reversed-phase
thin layer chromatography (RPTLC) on chemically bonded
phases already features over 90 references. For clarity's sake, it is emphasized here that RPTLC as such
can be subdivided into two categories--RPTLC on physically coated and on chemically bonded stationary
phases--and that it is only the latter type which will
be studied in the present paper.

In previous papers (2, 3) we have discussed the
relative merits of various types of commercially

87

available precoated plates for RPTLC, with emphasis on
aspects such as development time of the chromatogram
and compatibility of the RPTLC plates with mobile
phases containing a relatively large proportion of
water. Since then, several new types of commercial
RPTLC plates have become available. These are
included in the present study which, for the rest, has
mainly been directed at the separation of rather polar
compounds such as chloroanilines, chlorophenols, and
aminophenols.

        MATERIALS AND METHODS

        RPTLC Plates *

Commercial HPTLC-quality plates precoated with RP-8-
and RP-18-modified silica (HPTLC-Fertigplatten RP-8
F_{254s} and RP-18 F_{254s}, product no. 13725 and 13724,
respectively and TLC-quality plates precoated with
RP-8- and RP-18-modified silica (DC-Fertigplatten RP-8
$F_{254}S$ and RP-18 $F_{254}S$, product no. 15424 and 15423,
respectively, were obtained from Merck (Darmstadt,
GFR). The TLC-quality plates were received as gifts
in September 1980 (RP-8a), July 1981 (RP-18b), and
October 1981 (RP-8c and RP-18c), with the "a," "b,"
and "c" having been added by us for clarity.
        Precoated KC_{18} plates were purchased from Whatman
(Springfield Mill, U. K.; $KC_{18}F$ plates, product no.
4803-800). $Si-C_{18}F$ precoated plates were supplied by
Baker (Deventer, the Netherlands; product no. 7013-4),
OPTi-UP C_{12} precoated plates were obtained from Antec
(Bennwil, Switzerland; OPTI-UP C_{12} DC Traegerplatten
L 254; product no. 1028), and Nano-SIL C18-50 precoated plates were received from Macherey, Nagel & Co.
(Düren, GFR; product no. 811.064). All the above
20 x 20, 20 x 10, or 10 x 10 cm plates contained a
fluorescent indicator. They were used without pre-
treatment.
        TLC plates of a size appropriate for our experi-
mental work were obtained using the simple device
shown in Figure 1. Onto a metal or wooden base plate
a sheet of millimeter graph paper has been glued. A

* For further details on plate characteristics see
next section.

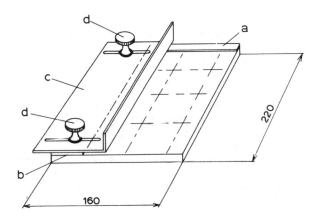

Figure 1. Device used for cutting TLC plates to
appropriate size. a and b, metal guides; c, metal
cutter guide; d, screws. Dimensions, 220 x 160 mm.
For further explanation, see text.

thin-layer plate is placed on this base plate, with
its precoated side down, so that it touches the metal
guides a and b. The metal cutter guide c is moved
into the desired position and secured by means of the
two screws d. Continuous scores are made in the glass
backing of the thin layer plate by moving a glass
cutter with its flat side along the edge of c. The
slides are then broken off at the edge of the base
plate. This procedure can conveniently be performed
with less than 5 percent loss of material.

Materials

The phthalate esters, aromatic acids, chloroanilines,
chlorophenols, aminophenols and further test solutes,
as well as all solvents and other chemicals, were of
normal analytical-grade quality. Approximately 1 percent sample solutions in dioxane were used for spotting. Dioxane was preferred to ethanol (cf. Ref. 2)
in order to prevent any esterification reactions with
the phthalate ester and aromatic acid stock solutions.

TLC

Chromatography was routinely carried out on rectangular plates of about 6.5 x 1.5 cm. After the application of about 0.5 to 1.0 mm spots using a pointed paper wick partly impregnated with the sample solution or a fine-tipped melting-point capillary, ascending development was done in Hellendahl staining jars over a distance of 5.0 cm. Saturation of the chromatographic chamber was achieved by lining its wider upper part with filter paper impregnated with the mobile phase. Visualization was accomplished by viewing the plates under 254-nm UV light. The aminophenols were detected as brown spots after 3 to 5 min irradiation of the RPTLC plates with a medium-pressure mercury lamp.

RESULTS AND DISCUSSION

RPTLC Plate Characteristics

A large variety of good-quality precoated plates for RPTLC is available, as is evident from the summary in Table I. Since most of these types of plates were used in the present work a few pertinent comments will be added (for the abbreviations used in the text, see Table I). The data given in this section have mainly been derived from product information bulletins or have been supplied by the manufacturers. For all references on the use of the RPTLC plates in analysis, one is referred to the review (1).

The Antec C_{18} plates have been available for several years, and they have been used in various studies. Recently, Antec appears to have changed the manufacturing process of these plates: newly received plates are much softer and migration times are considerably shorter; unfortunately, however, separation efficiency seems to have deteriorated. The average particle diameter, d_p, of the stationary phase material on the latter type of plates (which were mainly used in this study), is 10-20 (80 percent: 10:15) μm. The C_{12} plates use silica silanized with dodecyltrichlorosilane, with no subsequent end capping, and an inorganic binder. This may well explain why sorption of polar solutes such as

TABLE I. COMMERCIALLY AVAILABLE APOLAR CHEMICALLY BONDED RPTLC PLATES

Manufacturer	Plate Designation	Carbon Chain	Abbreviation in Text	F_{254} Indicated	Miscellaneous[a]
Antec	OPTi-UP C_{12}	12	C_{12}	+	Inorganic binder
Baker	Si-C18	18	—	±	± prescored
Macherey, Nagel & Co.	Nano-SIL C18- -100, -75, -50	18	MN-100/ 75/50	±	
Merck	Kieselgel silanisiert	2	—	±	
	RP-2,8,18	2,8	—	±	Acid-resistant F_{254}; TLC plates have improved water compatibility
	(HPTLC)	18		+	
	RP-8, -18 (TLC)	8,18	—	+	
Whatman	KC_{18}	18	—	±	End-capped with C2; ± preadsorbent strip

a ± With or without

91

nitrobenzene and acetophenone from apolar mobile
phases (cf. Ref. 4) in our experience was much
stronger with the Antec C_{12} plates as compared with
all other RPTLC plates tested.

The Baker Si-C18 plates are the most recent addition to the list of precoated RPTLC plates. As yet,
their use has not been described in the open literature, and data on plate characteristics are still
lacking.

Macherey, Nagel & Co., which markets a series of
three bonded-phase plates, use high-performance TLC-
grade silica (HPTLC silica; d_p stated to be 5-10 µm)
as base material. The silanol groups on this silica
are totally or partly (75 or 50 percent) reacted with
a C_{18} alkylsilane. No details are known about the
methods used to control and/or determine the degree of
silanization, and the real degree of silanization of
MN-75 and MN-50 materials is, by some workers,
assumed to be much higher, for example, on the order
of 90-95 percent for MN-75.

So far, the Merck RP-coated plates (RP-2, -8 and
-18) have been used most widely. Production started
some 5 years ago, the bonded phases being HPTLC (d_p ca.
5-7 µm) quality. With these precoated plates severe
problems arose, however, when mobile phases containing
over 30-40 percent of water were used. In such cases,
the nonwettability of the apolar stationary phase prevented normal chromatographic development; the addition of salts such as sodium or lithium chloride to
the mobile phase--a well-known remedy (2, 5) (cf.
below)--did not improve the situation. In order to
extend the application range of their precoated RPTLC
plates, Merck has recently marketed plates precoated
with C8- and C18-alkylmodified silica that feature a
slightly larger particle size of the silica used as
base material (d_p ca. 11-12 µm) while silanization is
not completely exhaustive. These are therefore called
chemically bonded TLC--as contrasted with the above
mentioned HPTLC--plates. They are essentially comparable with the well-known silanized silica precoated
plates manufactured since 1972 (Kieselgel silanisiert
F_{254}), which also feature a 11-12 µm particle size.
Here, it should be added that both these plates and
the RP-2-coated HPTLC plates are prepared via silanization with dimethyldichlorosilane, so that they are
(di-)RP-1- rather than RP-2-coated plates. Table I
lists the commercially available plates.

A survey of the literature reveals that an increasing number of workers prefer to use K_{18} plates (the KC_2 and KC_8 plates mentioned in a single recent paper are not commercially available). As is the case with the RP-coated HPTLC plates, the KC_{18} plates can not be used with mobile phases containing over about 40 percent of water. With the KC_{18} plates, however, it has been amply demonstrated that the addition of about 3 percent of sodium chloride to the mobile phase suffices to extend the application range to at least 80-90 percent of water in methanol-water and acetonitrile-water mixtures without any undue increase in time of development. For the rest, the manufacturer has repeatedly stated that the silica base material of the KC_{18} plates has a particle size of about 10 μm. Rather surprisingly, in a single recent brochure, 20 μm is mentioned instead; this seems to fit in rather well with provisional results from particle-size measurements (6).

Finally, the following characteristics of the RPTLC plates should be mentioned. When plates containing a fluorescent indicator are viewed under 254-nm UV light, the KC_{18} and Si-C18 plates display a strong, and the C_{12} plates a weaker, yellow fluorescent background; the (TLC) RP- and MN-type plates show a weak blue fluorescence. The KC_{18} and, slightly less so, the Si-C18 plates have a rather hard surface which, especially in the base of the Baker plates, has a fairly coarse structure. The RP- and MN-type layers are distinctly softer materials, which display a very smooth surface. With neither of these four types of plates does writing on the layer surface pose a problem. The C_{12} layers are smooth and extremely soft; small amounts of stationary phase material easily come off upon touching them, and writing cannot conveniently be done.

RPTLC Plate Quality

Model Separations. A rapid test for the assessment of the separation efficiency of the various types of precoated plates is provided by, for example, the analysis of a mixture of phthalate esters using a suitable methanol-water mixture as mobile phase. Results for seven types of plates--with emphasis on the RP-coated HPTLC and TLC plates--is shown in Figure 2.

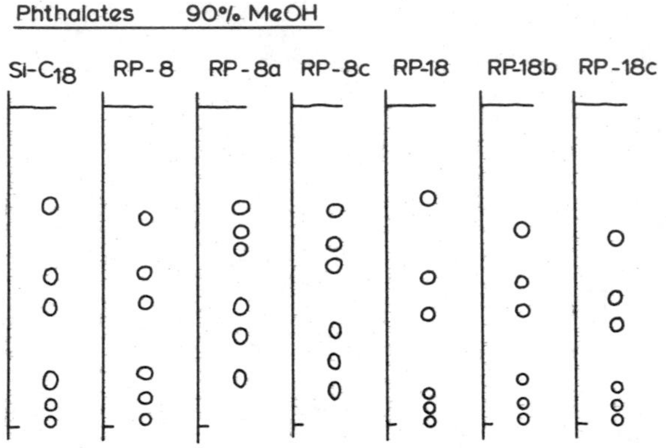

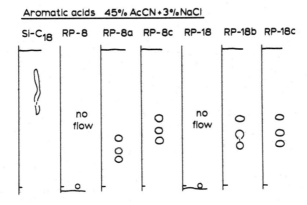

Figure 2. RPTLC, on Si-C18 and various types of RP-
coated plates, of six phthalate esters using methanol-
water (90:10) as mobile phase. Phthalate esters, in
order of decreasing hR_F: di-C_2, di-C_4, di-C_5, di-C_8,
di-C_{10}, di-C_{13}.

Figure 3. RPTLC, on Si-C18 and various types of RP-
coated plates, of three aromatic acids using acetonitrile-water (45:55) + 3 percent NaCl as mobile phase.
Acids, in order of increasing hR_F: p-nitrobenzoic,
acetylsalicylic, p-hydroxybenzoic acid.

Combination of these data with those previously published (2, 3) for, for example, KC_{18} and MN plates,
demonstrates that all precoated plates tested so far
pass the said test.
     The above test, in other words, is too simple to
be really valuable and, in its stead, a separation of
three aromatic acids was selected that requires the
use of a much more polar mobile phase (Figure 3). The
differences between the two types of separation are
obvious. In Figure 3, with SI-C18 development of the
plate poses no problem, but the separation is poor.
With the RP-8- and RP-18-coated HPTLC plates development does not occur at all, which is in agreement with
earlier results. The introduction of the RP-8- and
RP-18-coated TLC plates is seen to effect considerable
improvement: the flow of the mobile phase now is
uninterrupted, and an adequate separation of o-nitro-
benzoic, acetylsalicylic and p-hydroxybenzoic acid (in
order of increasing hR_F) is observed in all cases.
From previous work it is known (2, 3) that efficient
separations of the aromatic acids can also be created
on KC_{18} and MN-50 layers.
     With the C_{12} plates (data not shown), the results
of the phthalate ester separation were satisfactory,
though it is interesting that optimum separation was
achieved at 80 percent methanol as against 90-95 percent for all other RPTLC plates. With the aromatic
acids, no separation worth mentioning was achieved
with any of the mobile phases tested.

Migration Time. The dependence of migration time, for
a 5-cm run, on the mobile phase composition was system-
atically studied for several types of precoated plates.
Methanol-water and acetonitrile-water mixtures were
routinely used, both with and without added sodium
chloride. Development was done in saturated Hellendahl jars, but without previous accommodation of the
precoated plates.
     In Figure 4 results for the two available batches
of RP-18-coated TLC plates and KC_{18} plates are compared. The data for the KC_{18} plates are in good agreement with those published in Ref. 2: when using over
about 40 percent of water in the mobile phase, addition of sodium chloride is required and, provided this
is done, rather fast migration rates are obtained up
to at least 80-90 percent of water. With the RP-18

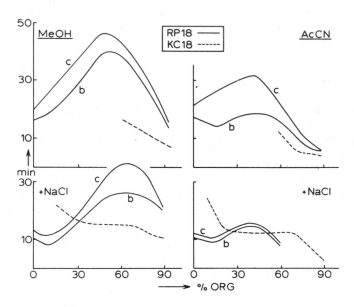

Figure 4. Dependence of migration time, for a 5-cm run, on the organic modifier (methanol or acetonitrile) and sodium chloride content of the mobile phase for two types of RP-18 (————) and for KC_{18} (-----) plates. No curve drawn for less than 60 percent modifier with KC_{18} without added NaCl because of very slow migration and/or TLC plate damage.

b and c normal development of the plates occurs at all mobile phase compositions (including 100 percent water) even without added sodium chloride. (Such an addition, however, generally increases the speed of migration to some extent.) Reproducibility of the results between batches is seen to vary between fair and excellent. For the rest, it is noteworthy that the shape of the migration time vs. mobile phase composition plots is completely different for the RP-coated TLC plates and the KC_{18} plates. With the latter type, the time of development is virtually independent of the mobile phase composition between about 30 and 70 percent of organic modifier (in the presence of sodium chloride), whereas there is a distinct maximum in the plots for the RP-coated plates. This more or less coincides

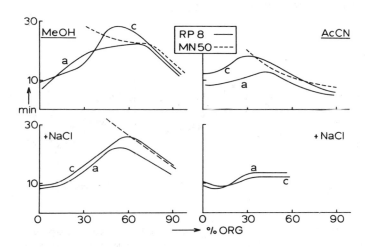

Figure 5. Dependence of migration time, for a 5-cm run, on the organic modifier (methanol or acetonitrile) and sodium chloride content of the mobile phase for two types of TLC-quality RP-8 (———) and for MN-50 (-----) plates.

with the maximum in the viscosity vs. composition plot of the mobile phase mixtures used.

Data for the two available batches of RP-8-coated TLC plates and for MN-50 plates are shown in Figure 5. With the RP-8-type plates, salt addition is again seen to be superfluous; reproducibility between batches is excellent in all four cases shown, and migration times are distinctly shorter than with the RP-18-coated plates. As for the MN-50 plates, with down to about 30 percent of organic modifier, their behavior is closely analogous to that of the RP-8-type plates. If, however, over 70 percent of water is present in the mobile phase, migration times increase rapidly--with the addition of salt detracting from rather than improving the performance of the system--and the plates cannot be used conveniently any more.

The C_{12} plates display very favorable migration characteristics. Migration times for the 5-cm run do not exceed 15 min over the whole 0:100 to 95:5 range for methanol-water and acetonitrile-water mixture.

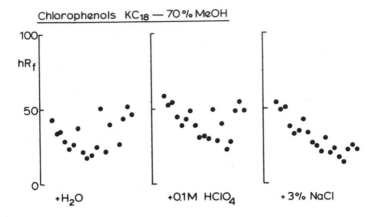

Figure 6. RPTLC, on KC_{18} plates, of 19 chlorophenols using methanol-water (70:30) (left), with added acid (center) or added sodium chloride (right) as mobile phase. Sequence of chlorophenols, from left to right: 2-, 3-, 4-, 2,3-, 2,4-, 2,5-, 2,6-, 3,4-, 3,5-; 2,3,4-, 2,3,5-, 2,3,6-, 2,4,5-, 2,4,6-, 3,4,5-; 2,3,4,6-, 2,3,5,6-; penta.

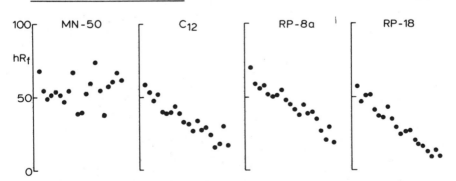

Figure 7. RPTLC, on MN-50, C_{12}, RP-8a and RP-18 (HPTLC) plates, of phenol and 19 chlorophenols (for sequence, see Figure 6) using methanol-water (70:30) as mobile phase.

Chlorophenols   MN-50 — 70% MeOH

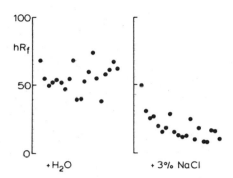

Figure 8. RPTLC, on MN-50 plates, of phenol and 19 chlorophenols (for sequence, see Figure 6) using methanol-water (70:30),without and with added sodium chloride, as mobile phase.

Addition of sodium chloride does not significantly influence the speed of migration. The behavior of the Si-C18 plates is closely analogous to that of the KC_{18} plates, shown in Figure 4.

On the basis of the above data, all further work was limited to RPTLC of rather polar test solutes since, with these, the effect of high water content of the mobile phase, and pH or ionic strength (sodium chloride) variation on chromatographic behavior can be more easily studied.

Chlorophenols

From reversed-phase column LC with, for example, methanol-water mixtures as mobile phase, it is known (7, 8) that for lower (mono to tri)chlorinated phenols the capacity factor, k', increases with an increasing number of chlorine substituents. With the tetra- and pentachlorophenol(s), however, such a relationship does not hold: retention starts to decrease and becomes irreproducible, while peak shapes tend to be poor. This anomalous behavior is caused by the rela- tively strong acidity of the higher chlorinated phe- nols, which are partly ionized under the experimental

Chlorophenols : LC vs. TLC

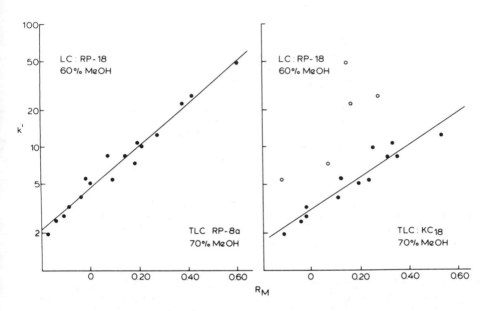

Figure 9. Plots of k' from RPLC on LiChrosorb RP-18 using acidified methanol-water (60:40) as mobile phase (Ref. 7) vs. R_M from RPTLC on RP-8a and KC_{18} plates using methanol-water (70:30) as mobile phase. Test compounds: chlorophenols; open circles denote outliers.

conditions. Acidification of the water used as (part of) the mobile phase to a pH of about 3 completely suppresses ionization, and capacity factors now increase almost linearly with increasing chlorine content through to the pentachlorophenol.

In the present work all 19 chlorophenols were chromatographed on KC_{18} plates with methanol-water (70:30, v/v) as mobile phase (Figure 6; in this and the subsequent figures 7, 8, 10, and 11 the chlorine content of the test solutes invariably increases from left to right). With methanol-water, the result is according to expectation. The hR_F values decrease in going from the mono- to the trichlorophenols and, then,

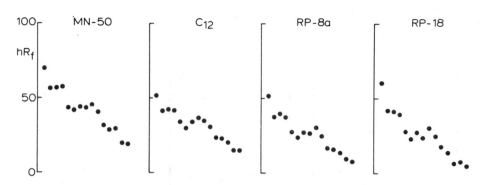

Figure 10. RPTLC, on MN-50, C_{12}, RP-8a, and RP-18
(HPTLC) plates, of aniline and 14 chloroanilines using
methanol-water (70:30) as mobile phase. Sequence of
chloroanilines: see Figure 6; 2,3,5-, 2,3,6-, and
3,4,5-tri-, 2,3,4,6-tetra- and pentachloroaniline not
included.

increase again with the higher chlorinated compounds.
Rather surprisingly, however, upon acidification of
the mobile phase (with 0.1M $HClO_4$), the general picture remains essentially the same. If, on the other
hand, some sodium chloride is added, the retention of
the tetra- and pentachlorophenol(s) distinctly
increases.

The unexpected results with the KC_{18} plates
prompted us to repeat this study with four more chemically bonded phases (Figure 7). Three of these, viz.
C_{12}, RP-8a, and RP-18 (HPTLC) yield the "normal" hR_F
sequence--that is, increasing retention from mono-
through to pentachlorophenol--even without acidification of the mobile phase. A rather muddled picture
emerges with the MN-5 plate. Here, results slightly
improved upon addition of sodium chloride to the
mobile phase (Figure 8) but, actually, the distinct
decrease of the hR_F values for all chlorophenols is a
much more striking phenomenon.

Finally, Figure 9 demonstrates the excellent
agreement between the column LC data and the RPTLC

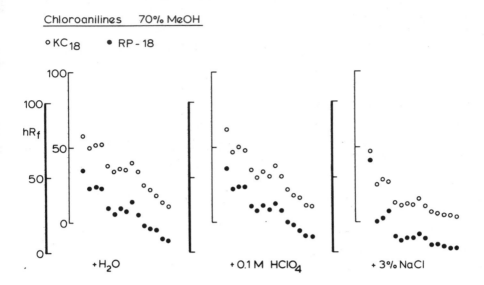

Figure 11. RPTLC, on KC_{18} (0) and RP-18 (HPTLC) (0)
plates, of aniline and 14 chloroanilines (for sequence,
see Figure 10) using methanol-water (70:30) (left),
with added acid (center) or added sodium chloride
(right) as mobile phase.

data obtained with the RP-8a plates. The results
clearly are less satisfactory in the case of RPTLC on
KC_{18} plates. Here, there are at least five outliers,
which are indicated in the figure by open circles.

   Chloroanilines

In order to find out whether the above results must be
attributed to the unique behaviour of the chlorophenols, a similar study was made with chloroanilines.
Here, all RPTLC plates tested show the same trend of
increasing retention with increasing chlorine content
of the substituted anilines, as is evident from
results for MN-50, C_{12}, RP-8a, and RP-18 (HPTLC) shown
in Figure 10. Besides, no changes occur upon the addition of either acid or sodium chloride to the mobile
phase as illustrated by the hR_F plots for KC_{18} and
RP-18 (HPTLC) in Figure 11.

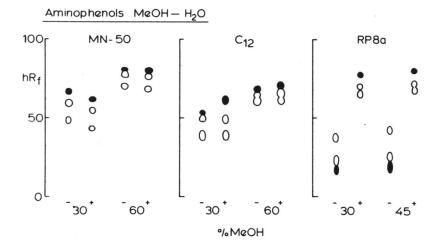

Figure 12. RPTLC, on MN-50, C_{18} and RP-8a plates, of para- (0), ortho-, and meta- (0; $hR_{F,ortho} < hR_{F,meta}$) aminophenol using various methanol-water mixtures (30, 45, or 60 percent methanol) without (-) or with (+) sodium chloride as mobile phase.

In agreement with the above results, plots of k' from column LC on LiChrosorb RP-18 vs. R_M from RPTLC on RP-8b and KC_{18} plates for mobile phases containing 50 or 60 percent of methanol showed straight-line relationships with correlation coefficients of over 0.985 in all cases.

    Aminophenols

The three isomeric aminophenols were selected as the last group of test solutes. They were separated on all types of plates available to us, using methanol-water mixtures--both with and without sodium chloride-- as mobile phase. In all but two cases, the hR_F values were found to increase in the order ortho < meta < para. Examples of this "normal" sequence are shown in Figure 12 for MN-50 and C_{12}, for various percentages of methanol.
    Exceptional behavior was observed for the RP-8- and RP-18-coated TLC (but not the RP-8- and RP-18-

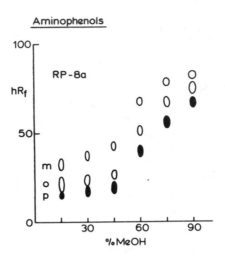

Figure 13. RPTLC, on RP-8 plates, of o-, m-, and p-
aminophenol using methanol-water mixtures as mobile
phase.

coated HPTLC) plates, and in the absence of sodium
chloride only (cf. Figure 12). With these, in the
presence of sodium chloride, the "normal" order of
separation was obtained. On leaving out the salt,
however, two striking changes occurred: (1) retention
increased considerably for all aminophenols, whereas
salt addition hardly affected hR_F values with the
other stationary phases tested; (2) with the para
isomer retention increased much more drastically than
it did with o- and m-aminophenol. As a consequence,
the order of separation was para < ortho < meta.
Similar results were obtained upon substituting acetonitrile for methanol. The order of separation did not
change when changing the organic modifier content of
the mobile phase; pertinent data for the RP-8a plates
are shown in Figure 13.
        Experiments similar to the above were carried out
for o-, m-, and p-nitroaniline, o-, m-, and p-chloro-
phenol, and o- and p-nitrophenol on RP-8a-, RP-18b-,
and KC_{18} plates. With these classes of compounds, and
with methanol-water (60:40) and (75:25) as mobile
phase, the order of elution was invariably the same,

both with and without added salt. In other words, a
relative increase in the retention of the para isomer
in the absence of sodium chloride was never observed.

CONCLUSION

Today, a large variety of thin layer plates precoated
with chemically bonded stationary phases is commercially available. The majority of these appears to be
of normal TLC quality (d_p > ca. 10 µm, the exceptions
being the RP-coated HPTLC-quality plates (d_p = 5-7 µm)
and, possibly, the MN-type layers (d_p = 5-10 µm).
Satisfactory allover separation efficiency is combined
with compatibility with mobile phases containing up to
at least 80-90 percent of water for the newly introduced RP-8- and RP-18-coated TLC plates, and with the
KC_{18} plates provided about 3 percent of, for example,
sodium chloride, is added to the mobile phase. The
MN-50 plates can be used up to about 70 percent of
water. The Merck RP-coated HPTLC plates and the MN-75
and MN-100 plates can be used over a limited range of
mobile phase compositions (less than 30-40 percent of
water) only. With the Si-C18 (salt addition!) and the
C_{12} plates, migration does not pose a serious problem
at any mobile phase composition, but here resolution
of the test solutes often is less good than with the
other types of plates. For the other precoated plates
tested, the plots of migration time versus mobile
phase composition mutually are strikingly different--
RP (TLC)-KC_{18} and Si-C18-MN-C_{12})--which seems to indicate fundamental differences in the manufacturing
process.
     The advantage of the RP-18-coated TLC plates of
unlimited use without added sodium chloride is partly
offset by the rather shorter migration time observed
with the KC_{18} plates in the often used 30-70 percent
methanol range. The combined advantages can possibly
be enjoyed when using the Merck RP-8-coated TLC layers
(cf. Figures 4 and 5). Lastly, it is obvious that
replacing methanol by the, unfortunately, rather more
toxic acetonitrile will decrease the time of development substantially. In this respect it may be worth-
while to study a number of other organic modifiers
such as acetone and dioxane.

All types of precoated plates display identical separation sequences with test solutes such as the phthalate esters and the chloroanilines. Considerable, and often unexplainable, differences occur, however, in the separation of, for example, chloro- and amino-phenols. As will be evident from the discussion in the previous sections, no general conclusion can be reached here as regards the preferred type of RPTLC plate. To quote one example, the RP-8- and RP-18-coated TLC plates should be recommended on the basis of the data reported for the chlorophenols; however, with the aminophenols, these two types of plates produce anomalous results. It seems more appropriate to conclude that, from the point of view of RPTLC separation studies, the varied--if unpredictable--behavior of the commercially available precoated plates adds to the potential of the technique as a tool for creating separations.

When using RPTLC as a screening technique for a rapid preliminary evaluation of mobile phase systems suitable for use in column LC, on the other hand, one should proceed with caution. Problems are most apt to occur in studies on the RP(T)LC of polar solutes. Here, mutual differences in interaction with residual silanol groups and/or binders, different pH values, and effects caused by the addition of neutral salts or ion-pairing agents may well ruin the hoped for similarity between RPTLC and RPLC retention data.

SUMMARY

The characteristics of eight types of commercially available thin-layer plates precoated with nonpolar chemically bonded phases are compared. Phthalate esters, aromatic acids, chlorophenols, chloroanilines, and aminophenols are used as test compounds.

ACKNOWLEDGMENT

We wish to thank Antec (Bennwil, Switzerland), Baker (Deventer, the Netherlands), Macherey, Nagel & Co. (Düren, GFR), and Merck (Darmstadt, GFR) for supplying us with gifts of their precoated chemically bonded plates.

REFERENCES

1. U. A. Th. Brinkman and G. de Vries, J. High Res.
   Chromatogr. Chromatogr. Commun. 5, 476 (1982).
2. U. A. Th. Brinkman and G. de Vries, J. Chromatogr.
   192, 331 (1980).
3. U. A. Th. Brinkman, G. de Vries, and R. Cuperus,
   J. Chromatogr. 198, 421 (1980).
4. M. Ericsson and L. G. Blomberg, J. High Res. Chromatogr. Chromatogr. Commun. 3, 345 (1980).
5. G. Grassini-Strazza and M. Cristalli, J. Chromatogr.
   214, 209 (1981).
6. A. M. L. Pluym, Janssen Pharmaceutica, Beerse,
   Belgium, personal communication.
7. C. E. Werkhoven-Goewie, U. A. Th. Brinkman, and
   R. W. Frei, Anal. Chem. 53, 2072 (1981).
8. K. Ogan and E. Katz, Pittsburgh Conference, March
   1980, Atlantic City, N. J., U.S.A.

CHAPTER 9
Discovery of
New Compounds by
Thin Layer Chromatography

Satinder Ahuja

INTRODUCTION

Chromatography provides an excellent means of discover-
ing new compounds since compounds present even at
ultratrace levels can be resolved from related compounds. One approach entails separation of potential
compounds resulting as by-products, which cannot be
resolved by normal inefficient crystallization techniques. Many of the by-products frequently have physicochemical properties and carbon skeleton similar to
the parent compound with substituent(s) differing in
position or functionality. Since it is not possible
to theorize all by-products, some unusual compounds
can be isolated and characterized with this approach.
A more selective approach is based on changes brought
about in a chemical entity to evaluate its stability
with reactions such as hydrolysis, oxidation, or
photolysis. In this case, several theorized new and
old compounds are produced. An innovative chromatographer can resolve and characterize both theorized and
untheorized new compounds. Another interesting
approach depends upon characterization of various
degradation products produced in the matrixes used for
pharmaceutical products. The compounds thus produced

TABLE I. PREDNISONE ASSAY VALUES OF SAMPLES
STORED UNDER ACCELERATED CONDITIONS

Batch	Storage	TPTZ Assay, [a] mg/cap
A	36 months, 40° C	3.67, 3.68
B	36 months, 40° C	2.28, 2.21
C	36 months, 40° C	1.74, 1.73
D	36 months, 40° C	1.93, 1.94

[a]   Theoretical value = 1.25 mg/capsule.

can be resolved from others by chromatography, and
their structure determined by techniques such as elemental analysis, IR, NMR, or mass spectrometry. Discovery of some of the compounds in our laboratory by
this approach with thin layer chromatography is discussed below.

RESULTS AND DISCUSSION

Phenylbutazone

A noncommerical pharmaceutical product that contained
prednisone, and some antacid excipients, when subjected to high temperatures for a prolonged period of
time, gave high assay values for prednisone by tri-
phenyltetrazolium colorimetric method (Table I).
     Triphenyltetrazolium chloride (TPTZ) or blue
tetrazolium (BT) have been commonly used in pharmaceutical industry as colorimetric reagents for determination of steroids containing α-ketol group. These
methods are based on the quantitative reduction of
colorless tetrazolium salts to colored formazans by
the α-ketol group in the presence of a base such as
tetramethylammonium hydroxide. The high prednisone
assay values suggested that a compound(s) that can
reduce TPTZ was being produced under these stress conditions. Samples stored under accelerated conditions
were analyzed by thin layer chromatography on silica
gel G plate with the cyclohexane-chloroform-acetic

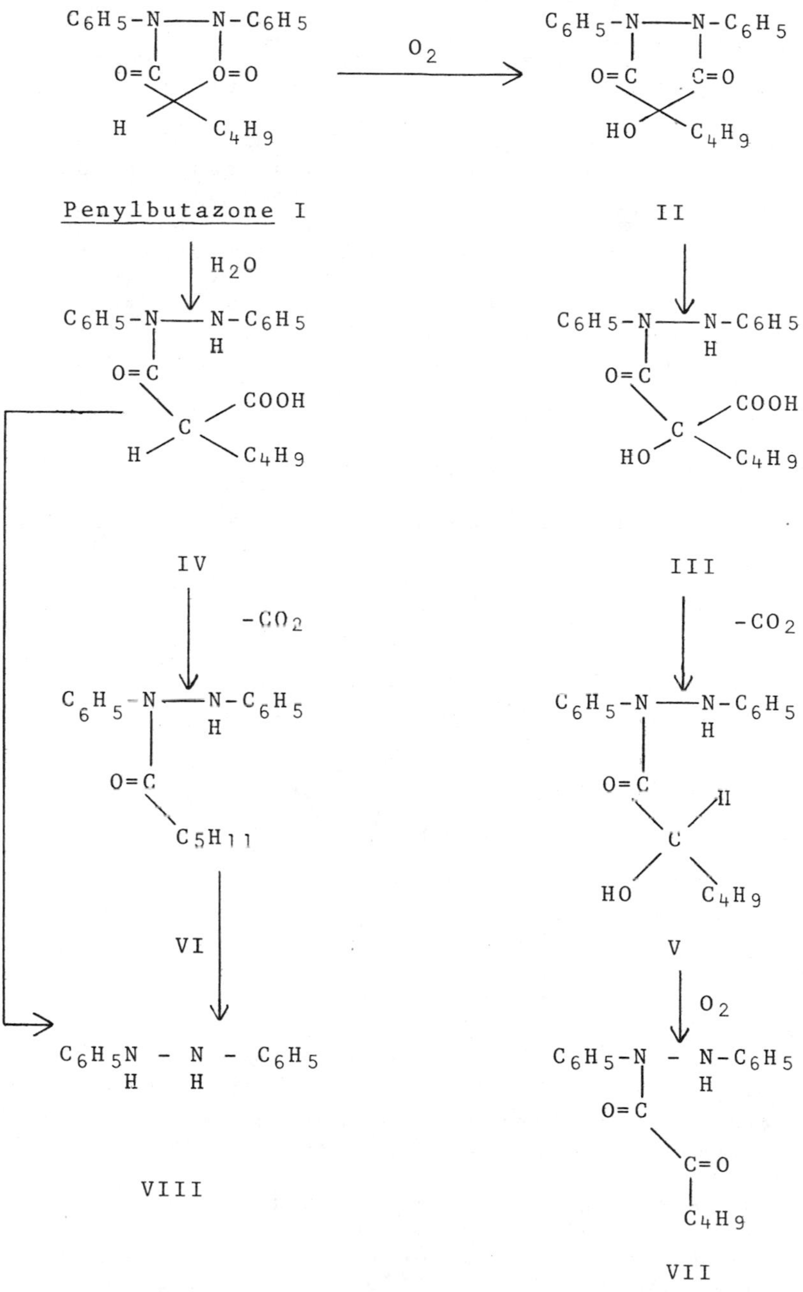

Scheme I. Phenylbutazone degradation.

Penylbutazone I

II

III

IV

V

VI

VII

VIII

111

acid (40:50:10) system.   When sprayed with $K_2Cr_2O_7/$
H_2SO_4 reagent, the presence of phenylbutazone (I) and
four impurities II, III, IV, and VI was revealed
(Scheme I).   These compounds do not produce color with
BT reagent.   Compounds such as hydrazobenzene, azoben-
zene, and caproic acid can be theorized as potential
impurities under extreme conditions.   Of these compounds, only hydrozobenzene exhibited reactivity with
BT.
      A blue tetrazolium spray reagent was developed in
our laboratory (1) wherein 0.35 percent blue tetrazolium solubilized in 95 percent ethanol is mixed in 2:1
ratio with 5N aqueous sodium hydroxide.   A TLC plate
was developed as described above and sprayed with BT
reagent.   A compound with the same R_f value as compound VI was found responsible for BT activity.   Since
hydrazobenzene has different R_f value than compound VI,
it appeared that a new compound(s) with the same R_f
value as compound VI was responsible for the high
assay values observed with prednisone.
      The spot that exhibited BT activity was isolated
by preparative thin layer chromatography.   The infrared spectrum of this compound (A) did not match that
of II, II, IV, and VI.   The following structure was
postulated based on the infrared spectrum:

Compound A (V)

The structure was confirmed by comparison of various
physicochemical data obtained with the authentic compound (2).
      Compound V gave poor BT activity.   This suggested
there were at least two new compounds present: one
with low BT activity (compound V) and the other with
high BT activity.   A new TLC system was developed to
resolve these compounds from the other compounds.   The

TABLE II. TLC OF PHENYLBUTAZONE AND ITS TRANSFORMATION
PRODUCTS WITH CYCLOHEXANE-CHLOROFORM-AMMONIUM
HYDROXIDE SYSTEM (40:50:10)

Compound	R_f
I, II, III, IV	0.0
A (V)	0.1
B (SA 484-82)	0.3
VI	0.4

TABLE III. ELEMENTAL ANALYSIS OF SA 494-82 (3)

Element	Theory, Percent	Found, Percent
Carbon	72.95	72.11
Hydrogen	6.80	6.59
Nitrogen	9.45	9.47

solvent system had virtually the same ratio of the
organic components except acetic acid is replaced with
ammonia. This results in a two-phase mixture--the
organic phase saturated with ammonia is used for development. It provides an excellent separation of the
desired components (Table II).

Compound B (SA 494-82) was isolated by preparative TLC. It has a melting point of 113° C. The
infrared spectrum indicates the presence of two car-
bonyl groups (1658 cm^{-1} and 1712 cm^{-1}) and NH function
(3300 cm^{-1}) (3).

Elemental analysis based on an empirical formula
$C_{18}H_{20}N_2O_2$ is shown in Table III.

The NMR spectrum was run in $CDCl_3$ and supports
the assigned structure VII (Table IV).

Further confirmation of the structure (VII) was
obtained with mass spectrometry (CEC-21-103C with
heated inlet at 180° C) data shown in Table V.

TABLE IV. NMR DATA ON SA 494-82 (3)

δ, ppm	Integration	Multiplicity	Assignment
0.8-2.0	7.3	Complex	($\underline{CH}$)
0.8	–	Asymmetric triplet	$-\underline{CH}$
2.6	1.8	Triplet	CH $-C=O$
6.5-7.7	9.9	Complex	Aromatic $\underline{CH}$

TABLE V. MASS SPECTRUM OF SA 494-82 (3)

m/e	Abundance, Percent	Assignment
77	73	$C_6H_5^+$
85	20	$CH_3(CH_2)_3$ $O=O^+$
91	38	$C_6H_5N^+$
92	18	$C_6H_5NH^+$
93	100	$C_6H_5NH_2^+$
106	14	$C_6H_5N_2H^+$
119	71	$C_6H_5NCO^+$
183	11	$(C_6H_5)_2N_2H^+$
296	(1)	M^+

The degradation pathway for formation of hydrazo-
bene (VIII) from phenylbutazone was established by
monitoring TLC of solutions of varying pH's (7.2-12.7)
maintained at 90° C. The results (Table VI) suggest
that hydrazobenzene can be produced directly from compound IV at higher pH (12.7) or its formation from
compound VI is extremely fast at this pH (4).

These investigations provided the basis for development of a method in which compound VII was selec-
tively extracted out. The prednisone assay values
thus determined with BT were uncompromised from phenylbutazone degradation.

Homatropine Methylbromide

Two-capsule formulations containing homatropine methyl-
bromide (HMB), antacids, and other ingredients showed
low HMB assay values for samples stored under accelerated conditions by the silver nitrate trimetric method.
Thin layer chromatography was explored with the acetic
acid-ethyl acetate-water-hydrochloric acid system
(35:55:10:20). All samples were spotted in 20 µl, and
the plates were developed to a distance of 10 cm.
After drying, the plates were sprayed with Dragendorff
reagent, and an orange-red spot was obtained for HMB.

In working with methanolic solutions of HMB it
was observed that HMB was unstable. A solution that
had been standing for 23 days at room temperature
gave a purple spot with Dragendorff reagent at R_f 0.25.
This is similar to the degradation spot observed in
capsules (5). The degradation compound could be conveniently isolated by refluxing HMB with methanol, followed by recrystallization with ether. This material
has a melting point of 344-347° C (decomp.)--litera-
ture (6): 346-347° C (decomp.)--and its IR spectrum
conforms to that found in the literature (6) for TMB.
Its purity based upon silver nitrate titration is 99.6
percent, and the results of microelemental analyses
are shown in Table VII. Based on these data, this
impurity was identified to be methyltropinium bromide
or tropine methyl bromide (TMB).

$$CH_2 \text{——} CH \text{——} CH_2$$

CH₂ —— CH —— CH₂
                N-CH₃   HC —— OH   ·   CH₃   Br
CH₂ —— CH —— CH₂

Tropine methyl bromide (TMB)
or methyltropinium bromide

To confirm that the impurity isolated from this
sample is the same as the one obtained on refluxing
HMB in methanol, efforts were made to isolate the
impurity by preparative TLC from one of the batches of
capsules stored at high temperature for a long time.
The infrared spectrum of the isolated material was
practically identical to the reference spectrum of TMB.
The small variations in the spectra were attributable
to the differences in the degree of purity and dryness
of two materials.

These investigations provided the basis for development of a method that is selective for HMB (5, 7);
TMB does not yield a precipitate with Dragendorff
reagent like HMB and thus does not interfere with the
assay.

CONCLUSIONS

Thin layer chromatography of noncommercial phenylbutazone formulations under accelerated conditions led to
discovery of two new degradation compounds of phenylbutazone. One of these compounds (VII) is responsible
for interference in prednisone assay with blue tetrazolium reagent. Similarly, TLC permitted characterization of a degradation product of homatropine methylbro-
mide in two noncommercial capsule formulations containing alkalizers. Based on these investigations, it was
possible to develop a selective assay methodology for
various dosage forms.

TABLE VI. DEGRADATION OF SODIUM PHENYLBUTAZONE (4)

pH	Compound	Time in Hours at 90° C					
		2	4	8	24	65	96
7.2	I	+	+	+	+	+	+
	IV	+	+	+	+	+	+
	VI	−	−	−	+	+	+
	Hydrazobenzene	−	−	−	−	−	+
8.0	I	+	+	+	+	+	+
	IV	+	+	+	+	+	+
	VI	−	−	−	+	+	+
	Hydrazobenzene	−	−	−	−	−	+
9.2	I	+	+	+	+	+	+
	IV	+	+	+	+	+	+
	VI	−	−	−	+	+	+
	Hydrazobenzene	−	−	−	−	−	+
10.2	I	+	+	+	+	+	+
	IV	+	+	+	+	+	+
	VI	−	−	−	+	+	+
	Hydrazobenzene						
12.7	I	+	+	+	+	+	+
	IV	+	+	+	+		+
	VI	−	−	−	−	−	−
	Hydrazobenzene	−	−	−	+	+	+

− = Not detected.

TABLE VII. ELEMENTAL ANALYSIS OF TMB

Element	Theory, Percent	Found, Percent	
Carbon	45.77	45.96,	45.72
Hydrogen	7.68	7.34,	7.35
Nitrogen	5.93	5.74,	5.70
Bromine	33.83	33.76,	34.02

REFERENCES

1. S. Ahuja and C. Spitzer, personal communication,
    October 27, 1966.
2. K. Adank, personal communication, October 27, 1966.
3. P. Nicolson, personal communication, June 13, 1967.
4. S. Ahuja and H. Burke, personal communication, July
    25, 1969.
5. D. Spiegel, S. Ahuja, and F. R. Brofazi, J. Pharm.
    Sci. 61, 1630 (1972).
6. K. Zeile and W. Schulz, Chem. Ber. 88, 1078 (1955).
7. S. Ahuja, D. Spiegel, and F. R. Brofazi, J. Pharm.
    Sci. 59, 417 (1970).

CHAPTER 10

New Methods in Analyzing Radiochromatograms and Electropherograms

Heinz Filthuth

INTRODUCTION

Two years ago I introduced at this Biennial Symposium
of Advances in Thin Layer Chromatography the Linear
Analyzer, a position-sensitive proportional counter or
wire chamber, to measure radiochromatograms and electropherograms. To remind you, the detector measures
the position of beta- and gamma-ray sources from a surface, like TLC plates or electrophoresis gels (Figure
1). The emitted beta-particles (or gamma-rays) produce electrical signals at the counting wires.
    The signals are induced into a delay line, and
their propagation time in the delay line is proportional to the position of the radiating source in the
TLC plate or gel which is positioned below the detector. Each particle entering the diaphragm of the
counter is detected and recorded, the propagation time
of the induced signal in the delay line is measured,
and therefore, each particle gets a label of its position and its angle of emission (inclination). The
last measurement electronically allows the selection
of the acceptance angle and, therefore, controls the
spatial resolution of the measurement (electronic collimation).

119

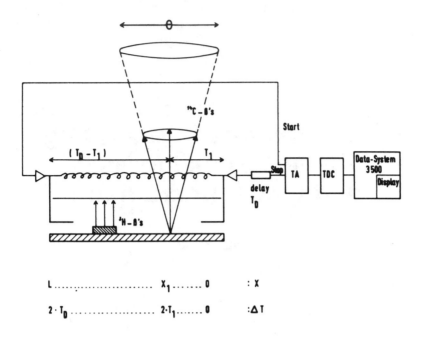

$$T_{start} = T_0 - T_1$$

$$T_{stop} = T_1 + T_0$$

$$\Delta T = T_{stop} - T_{start} = 2 \cdot T_1$$

$$T_0 = 500 \text{ nsec}$$

$$L = 250 \text{ mm}$$

$$0.25 \text{ mm} = 1 \text{ nsec}$$

Figure 1. Principle of position-sensitive proportional counter with delay line read out.

NEW DEVELOPMENTS

1. New Wire Chambers

We have recently developed two new wire chambers, named HR and HS, high resolution and high sensitivity.
With the HR-detector we obtain spatial resolutions of 0.5 mm for H-3 and 1 mm for C-14. With this detector it is also possible to measure high-energy beta-emitters and gama-emitters, like I-131, P-32, Tc-99m, and obtain spatial resolutions down to 1 mm.
To measure these isotopes one has to discriminate against the background radiation produced by the high-energy electrons in the chamber walls and the TLC plate. The background radiation deteriorates the spatial resolution.
We have limited the effect of background radiation in our new detector to a minimum.
The HS-detector is applicable for H-3 detection; it has about twice the detection efficiency of the former one for extended H-3 sources, that is, about 2 percent cpm/dpm for a 200-μm thick TLC plate.

2. Performance of the Linear Analyzer LB 283 with HR and HS Detector

The Linear Analyzer has the following properties:

1. Open entrance window 250 mm x 15 mm. For measurements of radioisotopes being not H-3 and I-125, the window can be closed by a thin foil 0.15 mg/cm^2. (In this case the resolution due to multiple scattering of the beta-particles is slightly worse.)
2. Background radiation:
   60 cpm/250 mm = 0.25 cpm/mm
3. Sensitivity:
   The detection efficiency for low energy beta-particles is practically 100 percent. But due to their absorption in the thin layer only a small fraction enters the counter. We have for a thin layer (HS-detector) of 20 mg/cm^2:

   H-3:   2.0 percent cpm/dpm
              (250 dpm detected in 20 min)
   C-14:   3 to 15 percent, depending on
              "electronic collimation"
              (50 dpm detected in 20 min)

Figure 2. Linear Analyzer LB 283 with Data System
LB 500 (Apple II computer).

| | P-32: | Up to 40 percent cpm/dpm |
| | Tc-99m: | 5 percent cpm/dpm |

4. <u>Resolution</u>:
     H-3:   0.5 mm (for a source covered with
                 1/2-mm plastic strip)
     C-14:   1.0 mm (for a source covered with 1-mm
                 plastic strip)
     P-32:   1.0 mm (for two strips on glass plate)

NEW DATA ACQUISITION SYSTEM

To "read" the detected signals the Linear Analyzer has
to be connected to a data acquisition system. Until
now we used a normal Multichannel Pulse Height Analyzer or the Data System 3500.
    Recently we also have developed a new data acqui-
sition system incorporating the well-known Apple II
computer (Figure 2). The signals from the preamplifi-
ers of the proportional counter are transformed into a
short (10 ns) start and stop signal (Figure 1). Since
the start-stop time difference is proportional to the
particle position, it is digitized and introduced into
a data system, like the Apple II computer. With the

PEAK SEARCH REPORT
(FULL ANALYSIS)

PEAK	CENTROID CM	LEFT CM	RIGHT CM	PEAK COUNT	FWHM CM	AREA	NET AREA
1	9.66	9.5064	9.7996	72	1.197E−1	4.240E2	3.955E2
2	10.72	10.459	10.948	52	2.017E−1	5.490E2	4.575E2
3	11.22	10.973	11.461	89	1.979E−1	8.520E2	7.493E2
4	12.80	12.610	12.952	66	1.402E−1	6.260E2	4.123E2
5	13.09	12.952	13.221	56	1.151E−1	5.100E2	2.997E2

PERCENT AREA SUMMARY

PEAK	PEAK AREA / AREA BETWEEN CURSORS	PEAK AREA / SUM OF PEAK AREAS	PEAK NET AREA / SUM OF NET AREAS
1	13.27	14.32	17.09
2	17.19	18.54	19.77
3	26.67	28.77	32.38
4	19.60	21.14	17.82
5	15.97	17.22	12.95

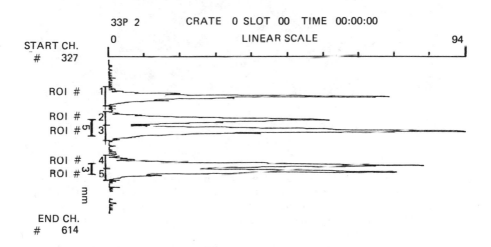

Figure 3. Hard copy of analysis report of P-32 measurement from 100-μm-thick HPTLC plate.

developed software one can measure automatically any
number of chromatogram tracks. For each track, one
uses the keyboard to type into the memory for that
track the title, preselected measuring time, or counts.
During measurements there is live display of the data

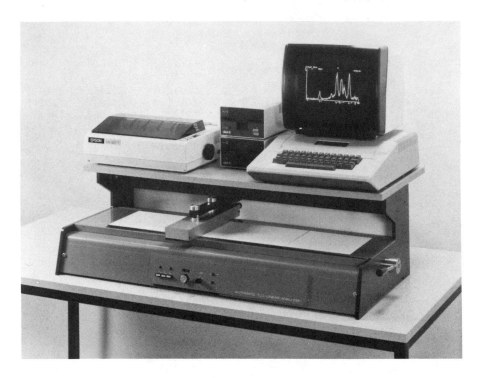

Figure 4. Multiplate Linear Analyzer with Data System
LB 500.

at the CRT screen. When the measurement of any track
is completed, the instrument stores the data in memory,
then sends a signal to position the next preselected
track below the diaphragm of the counter and initiates
a new measurement.

    After data acquisition has been completed for any
track, or for several tracks, the operator can use the
keyboard to recall data from memory. The instrument
displays the called (accumulated) data on the system's
CRT. The operator then can select the position scale
for measurement in any units, such as mm or R_f values.
He can demand a fully automatic analysis of the data.
The program finds all the peaks, subtracts background,
and gives on a hard copy a full report of the data
analysis, as peak position, width, counts, percentage
areas, and a graphics plot of the measured chromatogram, the ROIs being labeled (Figure 3).

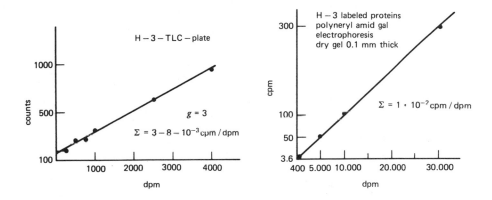

Figure 5. Sensitivity for H-3.

MULTIPLATE DETECTOR

To allow the automatic measurement of several chromatogram plates we have enlarged our present Linear Analyzer (Figure 4).

This new instrument is capable of measuring up to
four TLC plates, 20 cm x 20 cm, fully automatically.
The plates are fixed on a measuring table of about 1-m
length. They are pressed on two borders against two
thin steel bands from below. The counter glides under
computer control on the surface of these steel bands,
keeping always a fixed distance of 0.3 mm from the
plate surface. With the computer program, via keyboard and dialog, any number of chromatogram tracks to
be measured can be preselected, for example, their
position, the measuring time, or number of counts.

QUANTITATIVE MEASUREMENTS

The present system permits quantitative analysis of
beta-sources. The technique avoids the limitations
inherent in quantitation by using a scintillation
counter. For example, results for analysis of material labeled by H-3 and C-14 show that, within about
10 percent, the measured counts per minute are
directly proportional to the amount of radioactivity

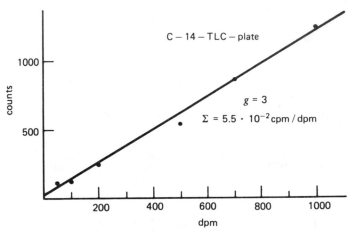

Figure 6. Sensitivity C-14.

introduced into the respective thin-layer gels (Figures 5, 6).   For the technique to work well, the
chromatogram should be uniformly distributed in depth
within the gel.   In this respect, very thin layers
result in more-precise, quantative measurement.   For a
quantative measurement, one has to consider several
important factors, which influence the result and certainly limit the precision.   Only beta-particles penetrating the surface of the thin layer or gel are
detected (Figure 7).   A particle originating at $Z = Z_0$
does not follow a straight line to the surface, $Z = 0$.
It collides with the atoms of the thin layer and therefore deviates from its original direction by the angle
Θ.   The mean multiple scattering angle is

$$\Theta = 15 / p.v \, (Z_0 / Z \, rad \,)^{1/2}$$

P is the momentum, v the velocity of the particle, and
pv is in MeV. $\cdot Z / \overline{Rad}$ is the radiation length of the
material. $\Theta \chi \, 30^\circ$ for a beta-particle of 100-KeV
energy traversing a 100-μm layer.
      The multiple scattering of the beta-particles
therefore limits the spatial resolution.   The beta-
particle loses energy before reaching the surface;
therefore, one detects beta-particles only from a

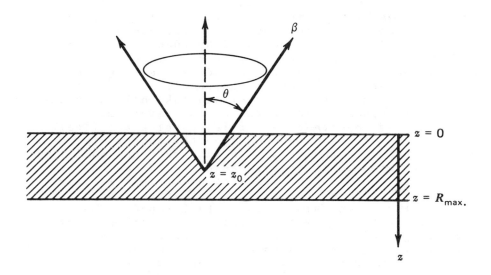

Figure 7. Emission of beta-particle from thin layer
plate. θ is the multiple scattering angle, R_{max} is
the maximum range of the beta-particle.

certain layer of thickness $R(max)$, corresponding to
its maximum energy in the beta-decay of the nuclid.
For example, H-3 has $R(max)$ = 0.55 mg/cm^2 - 8 μm layer
of thin layer plate, and already after R (1/2) = 0.06
mg/cm^2 = 0.8 μm, half of the emitted H-3-beta particles are absorbed and do not reach the surface.   Therefore, introducing a certain activity N (dpm) into the
thin layer, only $\Sigma_1 \cdot$ N gets to the surface and
$\Sigma_1 \cdot \Sigma_2$ N (dpm) are detected. Here Σ_1 is the absorption factor and Σ_2 is the detection efficiency of the
detector which is between 50 and 100 percent.
      From a 200-μm TLC plate we can detect in the very
best condition

$$\Sigma_1 = 8 \ \mu m/200 \ \mu m \cdot 2\pi/4\pi = 2 \ \text{percent}$$

of the H-3-decays, assuming that the H-3 labeled compounds have a uniform distribution in depth of the
thin layer and that the density of the thin layer does
not vary over the whole volume (X, Y, Z).   To deduce

from the measured counts (cpm) the final result (dpm),
we should know the distribution of the compound in X,
Y, and Z. This can be measured by running a chromatogram with known radioactive compounds in parallel to
the unknown chromatogram. To reduce the effect of a
variable depth distribution one should use very thin
layers, 100 µm thick and less, and take all the necessary precautions to generate the chromatogram.

     Another important factor is deducing the final
result, dpm of the compound, from the number of cpm is
the correct data analysis. To integrate the number of
counts related to the separated compound one has to
define its position and its contours. This is fairly
simple in the case of a single peak without background
or a uniform, nonrising, linear background. One can
integrate between the two border lines and subtract
the background measured in one or two sections near
the peak. All other cases are more complicated, and
the simple method I indicated is not sufficient to
obtain a precise result.

     If we would know the mathematical analytical
expression for the distribution of a chromatogram or
electropherogram, one could analyse the data with
standard χ^2 fitting methods.

     We have done this, assuming that the peak has a
gaussian shape and the background can be approximated
by a polynominal up to second order. This means the
chromatogram distribution can be described by the
expression

$$N(X) = \sum_i^2 dn\, X\, n + \sum_i h_i e^{-(x-x_i)^2 / 2\sigma_i^2}$$

where X_i is the position of the peak, h_i the height of
the peak, and $\Gamma_i = 2.35 \cdot \sigma_i$ the full width at half-
height. The area of the peak, A_i, is

$$A_i = \sqrt{2\pi} \cdot h_i \qquad \sigma_i \approx h_i\, \Gamma_i$$

     The χ^2 fit finds all the parameters a_n, h_i, i, X_i,
and allows then the correct calculation of the interesting quantities. An example of such a fit was
already shown above (Figure 3).

     This analytical method has the advantage of defin-
ing precisely the position, boundaries, and area of a
compound. It also is capable of identifying two

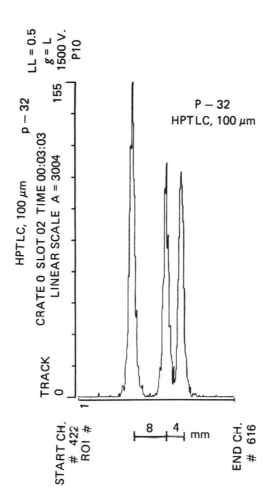

Figure 8. Detection of P-32. A resolution down to
1 mm is certainly possible.

overlapping compounds even if they are visually not
clearly separated. Instead of assuming a gaussian
peak one can of course approximate its shape by a different mathematical expression.
    The definition of background varies from experi-
ment to experiment. Background in our case is the
radiation not originating from the pure and, therefore,

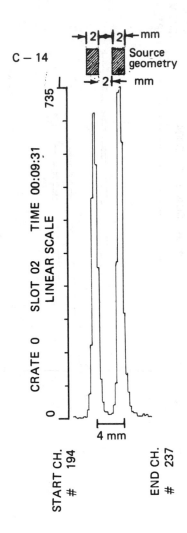

Figure 9. Detection of C-14 and P-33. The resolution
for both isotopes is the same. 1 mm can be obtained.

separated compound. The background is then composed
of signals from cosmic rays (1 cpm/cm^2), environmental
radiation, and signals from not separated, overlapping
compounds in the chromatogram. As long as one clearly

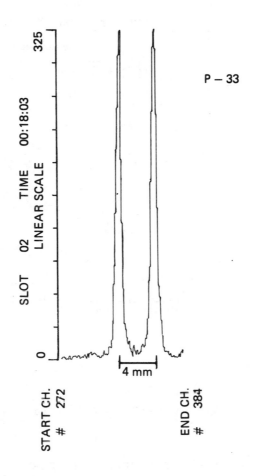

Figure 10. Detection of C-14 and P-33.   The resolution
for both isotopes is the same.   1 mm can be obtained.

defines the numbers extracted from a measurement
(chromatogram), there cannot be any confusion.   Consid-
ering all the factors I have described, it is certainly possible to achieve a precise measurement with
small, known systemic errors; the statistical error
can be reduced to practically any desired level.

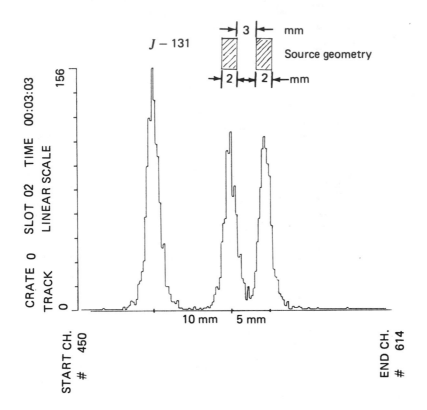

Figure 11. Detection of I-131.   A spatial resolution down to 1-2 mm is possible.

EXPERIENCE WITH THE LINEAR ANALYZER

Originally we have seen the application of the Linear Analyzer in detecting TLC chromatograms. In the mean-time, the application has been enlarged. We can detect radioactive labeled electropherograms in gels and radioactive distributions in organic tissues, and we can detect special nuclides like I-125, Tc-99m, P-32.

The following isotopes have been detected success-fully:

          Beta-emitters: H-3, I-125, I-131, C-14,
                           S-35, P-33, P-32
          Gamma-emitters: Tc-99m, I-123, Fe-59, Se-75

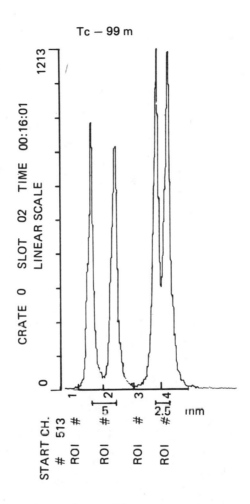

Figure 12. Detection of Tc-99m.   Total detection efficiency is more than 5 percent.   A spatial resolution
down to 1 mm is possible.

     The performance of the Linear Analyzer in measur-
ing the most common nuclides in TLC chromatography is
summarized in the earlier articles (1) and (2).   We
have improved our measuring technique and thus have
obtained better spatial resolution and increased sensitivity.

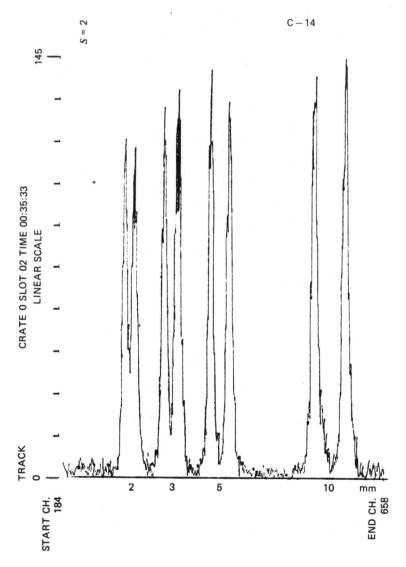

Figure 13. Measurement of C-14 labeled compounds in TLC plate.

TABLE I. TOTAL DETECTION EFFICIENCY, $\Sigma_1 \cdot \Sigma_2$, FROM THIN LAYER PLATE OF 200 μm THICKNESS WITH LB 283

Nuclid	Emitter	E(max)	Half-life	cpm/dpm, Percent
H-3	Beta	18 KeV	12 yr	2
I-125	Gamma-beta (auger)	30 KeV	60 days	2.5
I-131	Beta	608 KeV	8 days	Up to 20
C-14	Beta	155 KeV	5376 yr	Up to 20
S-35	Beta	200 KeV	87.5 days	Up to 20
P-32	Beta	1.7 MeV	14.3 days	Up to 40
P-33	Beta	200 KeV	25.3 days	Up to 20
Tc-99	Gamma	140 KeV	6 hr	5
I-123	Gamma	159 KeV	13 hr	5

Σ_1 is the fraction of beta-particles penetrating the surface of TLC-plate.
Σ_2 is the detection efficiency with LB 283.
Σ_2 ≤50 percent for H-3.
Σ_2 ≤90 percent for C-14, P-32.

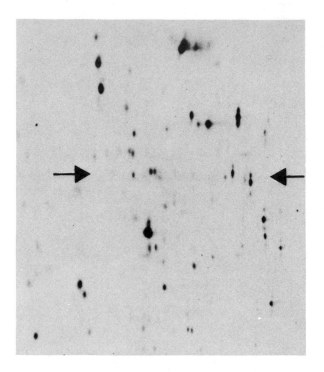

Figure 14 (1). Protein separation in two-dimensional
electrophoresis. Autoradiography.

     Table I summarizes the detection efficiencies of
the measured isotopes. In the case of I-125 we detect
the auger electrons and not the gamma-rays. The spatial resolution for I-125 detection is about the same
as for H-3, that is, better than 1 mm. The detection
efficiency from a 400-μm thin layer plate is 1.2 percent and from a 200-μm plate is 2.5 percent.
     Due to the high energy of the P-32 beta-particles,
E(max) = 1.7 MeV, it is difficult to detect them with
good spatial resolution. The energetic beta-particles
produce secondary electrons in the counter wall, in
the delay line, and in the thin layer plate or gel,
therefore, broadening the primary spots originating
from P-32. Also multiple scattering of the beta-
particles in the gel layer contributes to the broadening. This broadening effect we have reduced in our

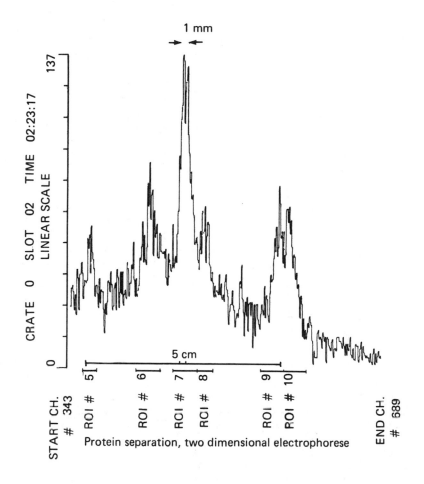

Figure 14 (2). Protein separation in two-dimensional
electrophoresis.   Detection with Linear Analyzer LB
283.

detector.   The same applies for the detection of gamma-
ray emitters, as Tc-99m and I-123.
    In the case of P-32 clear separation down to 3 mm
can be seen (Figure 8). Extrapolating these data, a
separation down to 1 mm is possible, the peak to valley ratio being about 3:2.   If the experiment demands
a higher resolution obtainable with P-32 labeled

compounds, it is recommended that P-33 be used, which
has a maximum beta-energy of 2-- KeV. For P-33 spatial resolutions similar to those for C-14 can be
obtained, as shown in Figures 9 and 10.

I-131 is a fairly high beta-energy nuclide,
E(max) = 608 KeV. Results of a measurement from a
TLC plate are shown in Figure 11. A spatial resolu-
tion down to 1 to 2 mm can be expected as for P-32.

Figure 12 shows the results for Tc-99m being a
pure gamma-ray emitter of 140-KeV energy. The two
peaks of 2.5-mm distance are clearly separated with a
peak to valley ratio of 3:1. For 1-mm separation a
ratio peak/valley of approximately 3:2 is expected.
The total detection efficiency is here 5 percent for
the high-resolution setting of the detector electronics (small).

I-123 is a gamma-emitter with gamma-energy similar to that of Tc-99m. We get results comparable to
those for Tc-99m.

Figure 13 shows a scan of C-14 labeled substances
separated on a thin layer chromatogram.

Figure 14 presents the results of a scan of a two-
dimensional electropherogram.

REFERENCES

1. H. Filthuth, New Detection Methods for Radio-
   Chromatograms and Electropherograms. In J. C.
   Touchstone (Ed.), Clinical and Environmental
   Applications of Quantitative Thin Layer Chromatography. Wiley, New York, 1980.
2. H. Filthuth, "Linear Analyzer Improves Detection in
   Radio-TLC Tests," printed in Industrial Research
   and Development, June 1981, pages 140-145.
3. H. Filthuth, "State of the Art in Scanning TLC:
   Detection of Radiochromatograms and Electropherograms with Position Sensitive Wire Chambers."
   International Symposium on the Synthesis and Applications of Isotopically Labeled Compounds, Kansas
   City, Missouri, June 1982.
4. H. Filthuth, Radioscanning of TLC. In J. C. Touchstone (Ed.), Advances in Thin Layer Chromatography.
   Wiley, New York, 1982, page 89.

CHAPTER 11
A Basic System for Quantitative Thin Layer Chromatography, Part I: Preparing the Chromatograms

J. Byron Sudbury
Ted T. Martin
Marvin C. Allen

In 1966, Killer and Amos published a formidable package of thin layer chromatography (TLC) procedures for the separation and identification of petroleum, petroleum products, and additives (1). Their publication clearly indicated the utility of TLC for qualitative analysis in the petroleum industry. In 1973, Martin expanded the scope of that message to promote the direct quantitative as well as the qualitative aspects of TLC for petroleum-related samples (2). At the December 1982 Advances in Thin Layer Chromatography Symposium, a basic TLC system in use at Conoco Inc. was described, and several examples of routine analytical methods were presented. Four chapters in this volume document this basic TLC system and two example methods.

This chapter discusses techniques employed in layer, track, and chromatogram preparation for quantitation. The next chapter covers the TLC densitometry system and the computer data handling procedures (3). Next follows a description of a method used to quantitatively analyze for compound types in source rock extracts (4). The final chapter presents a method used to analyze for compound types in high boiling fractions of refined petroleum (5). This series of chapters will hopefully provide insight into some of

Figure 1. Edge stripping. This figure illustrates
removal of a 9-mm strip of silica gel from two opposite edges of a layer.

the TLC techniques and methodology we currently use to
quantitatively characterize petroleum samples.

LAYER PREPARATION

Commercially prepared 20 x 20 cm TLC plates with 250-
μm soft silica gel G layers are used in a number of
routine quantitative analyses. The layers must be
free of bubbles, foreign matter, and any other visible
aberrations. Layers not satisfying these criteria are
either discarded or set aside for use in qualitative
applications.
        Prior to use, 9-mm strips of silica gel are
removed from two opposite edges of each layer (Figure
1). This promotes ease of plate handling and provides
a uniform margin of silica gel adjacent to the outer-
most samples during development. This "edge stripping"
is accomplished by using a razor blade mounted in an

Figure 2. Mounted razor blades used in edge stripping.
The blade extending 9 mm beyond the aluminum holder is
used to edge strip the two sides of the layer prior to
pre-elution, and the top of the layer after pre-elution.   The 4-mm blade is used to edge strip the bottom
of the layer following pre-elution.

aluminum holder, as illustrated by Stahl (6).   As the
blade is run along the surface of the plate, the
holder serves as a guide to remove a uniform amount of
silica gel.   When hard or thicker layers are used, the
silica gel can be shaved away instead of being removed
with one stroke.   The two mounted blades in use in our
laboratory are shown in Figure 2.
         After edge stripping, the glass plate backs are
cleaned and the plates are placed in a stainless steel
rack as shown in Figure 3.   The racks are used for
plate storage and are convenient for handling multiple
plates at one time.   The edge stripped zones are positioned so that the silica gel is not in contact with
the rods.   As shown in the schematic (Figure 4), each
rack consists of 1/8-in. stainless steel rods mounted
on 1/4-in. centers in a stainless steel frame.

Figure 3. Stainless steel plate rack. The edge-
stripped plates are placed in a stainless steel holding rack. Before and after pre-elution the plates are
stored in a clean tank such as the one illustrated.

        Prior to use in analyses, the layers are cleaned
to remove any organic contaminants present in the silica gel, a process we refer to as "pre-eluting" the
layers. A clean cylindrical "battery" jar (31 x 31 cm)
is used to pre-elute an entire rack of 28 plates at
one time (Figure 5). A glass plate is used in the jar
to provide a flat bottom. Pre-elution solvent (chloroform-ethyl ether-acetic acid, 50:50:4, v/v/v) is added
to a depth of 5.0 mm, the rack of plates is placed in
the jar and leveled, and the jar is covered with a lid

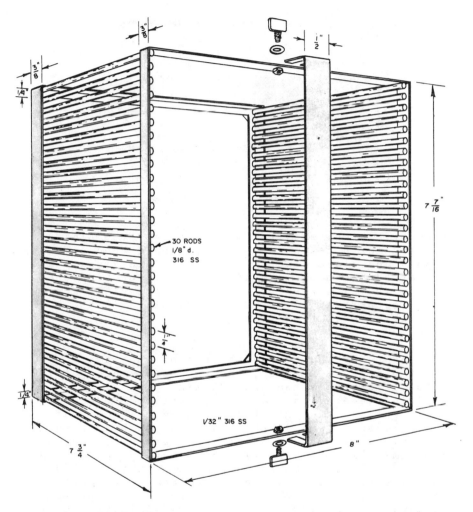

Figure 4. Stainless steel plate rack. Construction
details of the plate rack. The stainless steel rods
are welded in place.

(Figure 6). The pre-elution solvent requires about
60 min to move up the layers, and is permitted to
migrate for 80 min to make certain that the organic

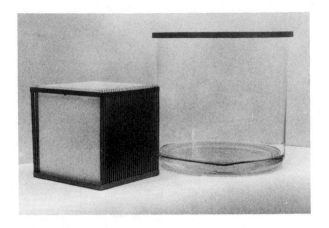

Figure 5. Plates ready for pre-elution. The plates
are ready for pre-elution once they have been edge-
stripped and placed in the rack. A clean glass plate
is inserted in the twenty-ninth slot to protect the
layer on the twenty-eighth plate. Glass vials are
used to support and level the glass plate in the bottom of the pre-elution tank.

contaminants are pushed to the top of the layer. The
rack of plates is removed, and the layers are permitted to air-dry in a hood for 20 min.
     This pre-elution process is repeated three more
times. Teflon™ washers of increasing thickness are
used to measure and adjust solvent depth before each
pre-elution. The four washers used are 5.0, 6.5, 8.0,
and 9.5 mm thick (Figure 7). By slightly increasing
solvent depth for each succeeding pre-elution, layer
impurities below the solvent level do not recontami-
nate the layer above the solvent level. The presence
of organic impurities in new commercial layers can be
readily confirmed by charring a pre-eluted layer.
     After the fourth pre-elution, the layers are
dried in a hood for 45 min. The entire rack of plates
is then placed in another cylindrical jar with an
aluminum lid for storage until the plates are needed
for analyses. Prior to use for analyses, the top 9 mm
of the pre-eluted layer is removed. This 9-mm zone
contains all of the organic material that the

Figure 6. Plates during pre-elution. After the rack
of plates is placed in the pre-elution tank, the rack
is leveled, and the aluminum lid promptly replaced.
Due to the flammable nature of the pre-elution solvent,
pre-elution and layer drying must be carried out in a
hood away from ignition sources.

pre-elution solvent pushed to the top of the layer.
The bottom 4 mm of the layer is removed to prevent solvent from wicking onto the layer during equilibration.
Both zones are removed with the blades previously
illustrated (Figure 2).

TRACK PREPARATION

Tracks (lanes) are an integral part of our system of
quantitation via fixed slit densitometry. The narrow
tracks serve to restrain sample spread during development and permit more samples to be analyzed per plate
than would otherwise be possible. As described elsewhere, these tracks are also essential to our system
of automated densitometry (3).
        Immediately prior to use, each layer is subdi-
vided into 25 tracks. A schematic of the tracking

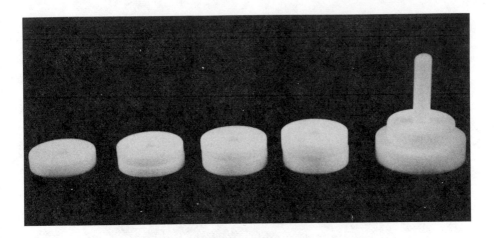

Figure 7. Teflon™ washers. A series of Teflon™ washers are used as depth gauges to measure pre-elution
solvent level. More solvent is used for each succeeding pre-elution to avoid recontamination of the layer.
The washers illustrated are (from left to right) 5.0,
6.5, 8.0, and 9.5 mm thick. For comparative purposes,
the depth gauge (12 mm) from the first developing tank
is shown at the far right.

device is shown in Figure 8. The actual tracking
device is pictured in Figure 9. The grooves in the
layer are made by drawing a bar, containing 26 spring-
loaded center punch points mounted on 1/4-in. centers,
the length of the plate. The bar runs on a fixed
shaft so that the tracks will be straight. The plate
fits into a template to assure reproducible track
location from plate to plate. A vacuum drawn on the
plate back prevents movement of the plate during tracking. After tracking, an air blower is used to remove
silica gel dust.
        Figure 10 illustrates the product of our tracking
device: 25 uniform tracks approximately 5 mm wide for
chromatogram development. At this point, the plate is
densitometered to record background density readings
for each track; the mechanics of this fully automated
process are described in more detail elsewhere (3).

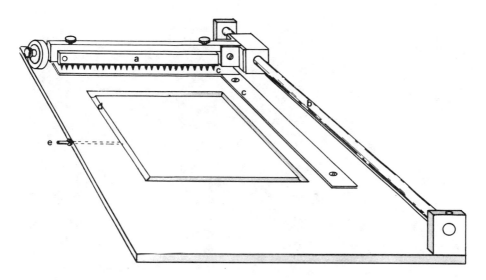

Figure 8. Tracking device. This schematic graphically
illustrates the major components of our tracking
device: a. Bar with center punch points.   b. Fixed
shaft.   c. Template for corner of plate.   d. Vacuum
trough which draws against plate back.   e. Outlet to
vacuum line.

SAMPLE APPLICATION

After tracking and densitometry of the blank tracks,
the plate is placed on a glass template (Figure 11).
A line on the template, 28 mm above the bottom of the
plate, marks the location for sample application. The
other lines on this template denote the desired distance for solvent migration in various methods and may
be marked with India ink on the edge-stripped portion
of the plate. The template also has a number scale
across the bottom identifying the 25 tracks. Samples
are applied to the appropriate tracks using disposable
glass capillaries of the desired capacity.

Some years ago, we began permanently mounting the
silicone rubber tip of the commercially supplied
capillary holder in a pair of forceps (Figure 12).
This is a very convenient tool for holding the capillaries during sample application (Figure 13). The

Figure 9. Tracking device. This is the actual tracking device with a plate in place. In a, the bar is
pivoted on the shaft to exhibit the 26 center punch
points. The TLC plate is held in place by vacuum.
Photos b-d illustrate the actual process of making
grooves in the layer.

desired volume of each standard and sample solution is
transferred to the appropriate track via capillary.
In sample transfer, the sample aliquot may be taken up
in one end of the capillary and dispensed from the
other end. This procedure avoids possible transfer of
sample from the outside surface of the capillary.
     After sample and standard application, the layer
is ready for processing. In our sample log-in procedure, each plate is assigned a successive number and
the sample or standard to be applied to each track is
recorded in the sample log book. The plate number is
written with India ink at the top of each plate on the
edge stripped zone. When the final visualized chromatograms are photographed, the plate number provides
a ready reference to the samples analyzed on a given
plate.

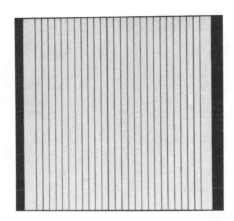

Figure 10. Tracked TLC plate. Following tracking, the
pre-eluted edge-stripped plate has 25 tracks available
for standard and sample chromatogram preparation.

EQUILIBRATION AND DEVELOPMENT

Most of our routine chromatograms are prepared following equilibration in a developing tank with an atmosphere saturated with developing solvent vapor.   A
7-3/4 x 24 in. piece of filter paper is folded to fit
the inside wall dimension of a standard-design TLC
developing tank (Figure 14).   This filter paper serves
as a liner which helps to saturate the tank atmosphere
with solvent vapor prior to and during layer equilibration and development.   The two liner ends may be
attached with staples placed above the solvent line.
In the case of extremely corrosive solvents, paper
clips for securing the two ends of the liner may be
made of stainless steel.   New tank liners remain in
solvent overnight to permit them to pre-elute, and the
solvent is replaced prior to chromatogram development.
    After the pre-eluted liner is installed in the tank, the
desired developing solvent is added to the tank at a depth
greater than the pre-elution solvent (9.5 mm) and short of the
sample application area (28 mm). A depth gauge made of Teflon™
is used to help evaluate the proper amount of solvent to add;
the initial solvent depth routinely used is 12 mm. When multiple developments are required, each succeeding developing
solvent is 1 to 3 mm deeper.

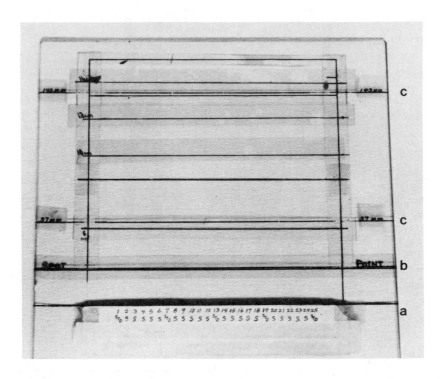

Figure 11. Glass template. Prior to sample application, the tracked TLC plate is placed on this template
with a number scale which indicates the number of each
of the 25 tracks (a), which marks the area 28 mm above
the plate bottom for sample applications (b), and
which indicates maximum extent of developing solvent
migration (c).

Before placing a plate in the developing tank for
equilibration, the solvent is permitted to wick up the
tank liner so that the tank atmosphere will be saturated. For equilibration, the tank with liner and
developing solvent is placed in a 9 x 9 x 7 in. card-
board box and leaned forward at approximately a 30°
angle (Figure 15). The tank lid is removed, and the
bottom edge of the TLC plate is placed along the
drained back of the tank bottom. The top edge of the
TLC plate is leaned forward against the front of the

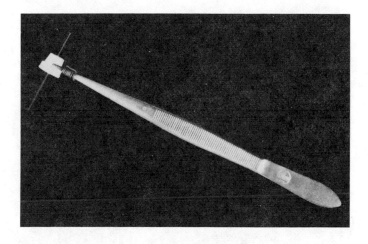

Figure 12. Capillary holder. Forceps with the tips
permanently wrapped around a silicone rubber capillary
holder tip make an excellent tool for holding glass
capillaries while applying samples.

Figure 13. Capillary holder. Here the capillary
holder is being used to dispense 3 µl of standard
solution to track number 7.

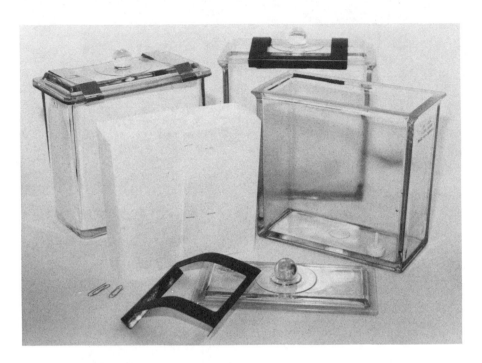

Figure 14. TLC developing tanks, liners, and clamps.
This figure illustrates a number of details about our
developing tanks, liners, and clamps. Filter paper
liners are made to fit the inside wall dimension of
the developing tank. Liner ends may be stapled, or
connected by paper clips (foreground). The 12-mm Teflon ™ depth gauge used for the first developing solvent is shown in the empty tank. Two varieties of
clamps used to secure the tank lid to the tank are
also illustrated.

tank, with the layer side down. Two spacers made of
Teflon™ are attached to the top edge of the plate to
keep the layer from contacting the liner (Figures 16
and 17). Clamps or spring clips are used to secure
the lid to the tank (Figures 15 and 17). As the plate
is quickly installed after the tank is saturated, this
is an effective means of layer equilibration. A typical layer equilibration period lasts for 15 to 20 minutes.

Figure 15. Developing tanks with liners. The layer in
the left tank is equilibrating with developing solvent
vapor, and the chromatograms on the tracks in the
other tank are developing.

     To terminate the equilibration and begin the
development, the developing tank is gently returned to
the upright position. As TLC data acquisition and
computer data processing are automated, visualized
spots must fall within previously determined windows.
Experience with various routine analyses has enabled
us to determine how far the solvent front must travel
to place the sample spots in the desired windows.
Thus, development is terminated when the solvent front
reaches this point on the layer, denoted by the India
ink lines on the edge-stripped portion of the plate.

          SOLVENT REMOVAL FROM LAYER

Although the layer will dry on prolonged exposure to
air in a hood, a heat gun is normally used to acceler-
ate developing solvent evaporation. A routine visualization technique we use is char by exposure to SO_3
fumes. The overall efficiency of this particular technique is enhanced by a layer that is free from

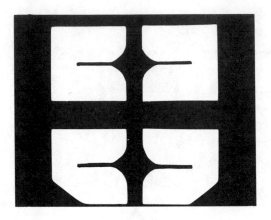

Figure 16. Teflon™ spacers. These two styles of Tef-
lon™ spacers are used to hold the top edge of the plate
away from the tank liner during equilibration and
development.

moisture that might be absorbed by the layer from the
air as well as from organic developing solvent vapors.
     For this reason, modified vacuum ovens are rou-
tinely used to dry out layers prior to visualization.
This particular oven has been previously described (7).
Basically, a commercial vacuum oven was modified by
use of a solenoid valve and a low pressure control to
enable the ovens to be operated under continuous gas
flow, continuous vacuum, or intermittent gas sweep/
vacuum (Figure 18). The inert gas or air entering the
vacuum oven can be drawn through a bed of charcoal or
desiccant and/or filtered. When these vacuum ovens
are used for drying layers, they are normally operated
in a filtered air-swept mode at 100° to 130° C and at
350-400 mbar (10-12 in. Hg) for 10 to 60 min, depending on solvent and sample type. After vacuum oven drying, if an additional development is desired, the
plate is placed in a desiccator to cool. (These same
drying and cooling procedures may be used if a layer
is to be activated prior to sample analysis. However,
prolonged clean activated layer storage in a desiccator is not recommended because of probable layer
recontamination with heavy organic impurities.)

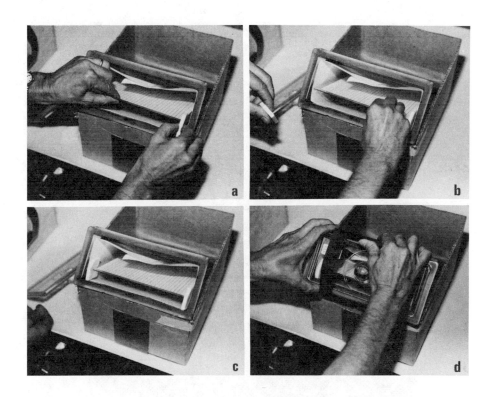

Figure 17. Teflon™ spacer placement. The lid is
removed from the tilted developing tank, and the bottom edge of the TLC plate is placed along the dry edge
of the tank bottom (a). A Teflon™ spacer is installed
at each edge of the plate on the edge-stripped zone
(a-c). The lid is then promptly put back on the tank,
and the clamp installed to secure the lid to the tank
(d). In this position, the layer can equilibrate with
the developing solvent as long as is necessary.

VISUALIZATION

Following development and drying, the chromatograms
must be visualized. Although there are literally hun-
dreds of visualization techniques and reagents available, we have found heating the plate in an atmosphere
of SO_3 to be the most effective visualization

Figure 18. Modified vacuum oven. The solenoid valve
and low-pressure control on this vacuum oven enable
operation under continuous gas flow, continuous vacuum,
or intermittent gas sweep/vacuum. Air entering the
vacuum oven during "sweep" operation may be dried by
passing through a column containing desiccant. For
safety and convenience, our glass char lids are normally stored on top of our vacuum ovens.

technique for the sample types described in later
chapters (4, 5). This char method is described elsewhere (8) and has been successfully used to visualize
a variety of sample types (e.g., 9, 10, 11).
        A 10-1/2 x 10-1/2 x 2 in. aluminum block is
mounted on a 9 x 9 in. hot plate in a well-ventilated

hood (Figure 19).   The thermostat is adjusted so that
the temperature in the center of the aluminum block is
150° C.   A 14 x 14 x 1/4 in. piece of Pyrex™ glass is
used to cover the aluminum block, and an 8 x 8 x 1/8
in. piece of Pyrex™ glass is placed on top of the cover
plate.   The resulting surface temperature of the 8 x 8
in. piece of glass is 125° C.   The aluminum block
helps to evenly distribute the heat to the glass and
to the TLC plate, which is taken from the vacuum oven
and placed directly on the 8 x 8 in. piece of Pyrex™
glass.

     A glass lid from a 9 x 9 in. Sunbeam™ electric
skillet is annealed, and the inside surface sand-
blasted to increase its surface area and acid-holding
capability.   A glass wool swab held in forceps is used
to apply the fuming sulfuric acid (30-33 percent SO_3)
to the glass lid.   Once the inside surface of the
"char lid" is coated with acid, the lid is placed over
the TLC plate (Figure 19), and the organic components
in the chromatograms are charred by exposure to the
SO_3 fumes.   No liquid acid touches the layer.

     Following a 20-min exposure of the layer to the
SO_3 fumes, the char lid is removed and the TLC plate
is left on the char block for several minutes to
enable the residual SO_3 fumes to bake out of the layer.
The TLC plate is then removed from the char block, and
the glass surfaces cleaned.   After the plate number is
reapplied to the edge-stripped zone, the plate is
stored in the dark until it is densitometered.

     The background densitometer values obtained for
the blank tracks prior to sample application will be
used to correct the charred track data for irregulari-
ties in the layer.   This topic, along with data
acquisition, computing, and reporting, will be covered
in the following chapter (3).

     CONCLUSIONS

With this system of chromatogram preparation, 25 sample and standard chromatograms are prepared and visualized on one plate.   The number of plates prepared per
day depends on the sample type being analyzed, as sample preparation, number of developments, and length of
drying times vary.   However in most of our routine
applications, one technician processes at least four

Figure 19. Fuming SO_3 char. A glass wool swab is
dipped into a beaker containing freshly poured H_2SO_4
(30 percent SO_3) and then briefly permitted to drain
(a).   The acid is then applied to the glass char lid
and evenly distributed by use of the glass swab (b,c).
The acid-coated lid is then inverted over the TLC
plate on the char block and the SO_3 vapor char permitted to continue as long as desired (d).

to six plates per day, and this includes sample log-in
and preparation, chromatogram preparation and visualization, data acquisition, computing, and reporting.
Thus, with this system, 100-150 chromatograms suitable
for accurate quantitation can be prepared and evaluated daily on a routine basis.

Teflon - Registered trademark of E. I. du Pont de
     Nemours and Company
Sunbeam - Registered trademark of Sunbeam Corporation
Pyrex - Registered trademark of Corning Glass Works

ACKNOWLEDGMENTS

For their contributions to the success of the TLC
effort at Conoco Inc., we wish to thank L. G. Becraft,
R. W. Bockhorst, C. T. Burch, P. E. Burnett, D. B.
Burrows, J. B. Coon, D. E. Cooper, R. K. Cormick,
L. A. Dye, H. H. Ferrell, O. F. Folmer Jr., W. C.
Hamilton, G. J. Hanggi, J. R. Harmon, D. Harrison, and
others.

REFERENCES

1. F. C. A. Killer and R. Amos, J. Inst. Petrol. $\underline{52}$,
   315 (1966).
2. T. T. Martin, Preprints, Div. of Petrol. Chem.,
   ACS, $\underline{18}$, 562 (1973).
3. T. T. Martin, M. C. Allen, and J. B. Sudbury, in
   Techniques and Applications of Thin Layer Chromatography, J. C. Touchstone and J. Sherma (Eds.),
   Wiley, New York, 1984, Chapter 12.
4. T. T. Martin, M. C. Allen, and J. B. Sudbury, in
   Techniques and Applications of Thin Layer Chromatography, J. C. Touchstone and J. Sherma (Eds.),
   Wiley, New York, 1984, Chapter 13.
5. J. B. Sudbury, T. T. Martin, and M. C. Allen, in
   Techniques and Applications of Thin Layer Chromatography, J. C. Touchstone and J. Sherma (Eds.),
   Wiley, New York, 1984, Chapter 14.
6. E. Stahl, Thin Layer Chromatography, A Laboratory
   Handbook, Academic Press, New York, 1965, page 11.
7. T. T. Martin, Lab. Pract. $\underline{21}$(7), 497 (1972).

8. T. T. Martin and M. C. Allen, JAOCS $\underline{48}$(11), 752
   (1971).
9. W. V. Allen, Comp. Biochem. Physiol. $\underline{47A}$, 1297
   (1974).
10. T. Wolf and B. P. McPherson, JAOCS $\underline{54}$, 347 (1977).
11. T. Jupille, J. Chromatogr. Sci. $\underline{17}$, 160 (1979).

CHAPTER 12

A Basic System for Quantitative Thin Layer Chromatography, Part II: Data Acquisition, Computing, and Reporting

Ted T. Martin
Marvin C. Allen
J. Byron Sudbury

INTRODUCTION

The preparation of large numbers of chromatograms for quantitation by thin layer chromatography (TLC) has been described (1). This chapter will show how the data acquisition, computing, and reporting were automated.

INITIAL INSTRUMENTATION

Figure 1 includes some equipment that was used in a 1967 version of the current quantitative TLC system. This initial instrumentation consisted of a Photovolt single-beam transmittance densitometer with multiplier photometer and a Varicord recorder (2).

The densitometer had a synchronous, motor-driven stage or plate carriage with focused light source below and photomultiplier tube detector above. A 4 x 0.5 mm masking slit was used for the source, and a 15 x 0.1 mm collimating slit was used for the detector. With this arrangement, 80 percent of the width of a 5-mm wide track could be automatically scanned along its length. The analog signal generated by the detector

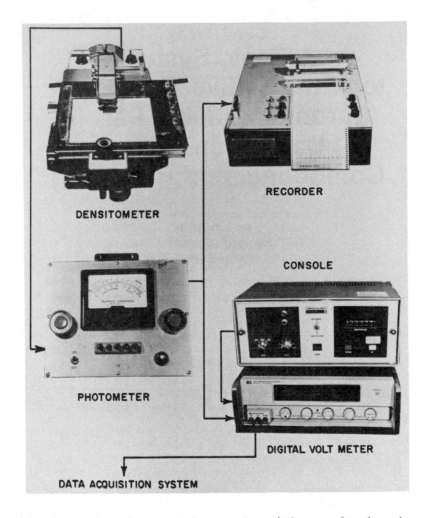

Figure 1. Initial instrumentation (Photovolt densitometer, multiplier photometer, and Varicord recorder)
employed in 1967. Interfacing instruments (Hewlett-
Packard digital voltmeter and Perkin-Elmer console)
temporarily added in 1970.

was amplified by the photometer. That signal was then
fed to the recorder and used to produce a density profile trace of the track on a strip chart. The

variable response feature of the Varicord recorder
could be used to obtain net peak areas that were
directly proportional to and linear functions of sample
component concentrations (3).

   The initial Photovolt instrumentation worked well.
However, as more analytical methods were developed and
the routine sample load increased, the efficiency of
the whole operation had to be improved. Innovations
that significantly increased the number of chromatograms that could be prepared only made the need for a
faster data acquisition, computing, and reporting system more urgent.

   Finding that the speed of the densitometer scan
motor could be doubled and quadrupled helped some, but
much more than that was needed. Each track still had
to be manually set up for automatic scanning, and processing the strip chart data was even more labor-
intensive and time-consuming. To help solve these
problems, an automatic track-changing apparatus was
designed. While this device was being built and
tested, the Photovolt instrumentation was temporarily
interfaced to an analytical data acquisition system
(ADAS) which had been installed at another location in
the same laboratory building.

   TEMPORARY INTERFACE

The temporary interface of the initial Photovolt
instrumentation to the ADAS was completed in 1970 as
indicated in Figure 1. A Hewlett-Packard digital
voltmeter (DVM) was connected in parallel with the
recorder. The DVM converted the amplified detector
signal from analog to digital form and sent it to the
ADAS, where it was recorded on magnetic tape. The
Perkin-Elmer digital readout console was used to
select a data rate, activate the ADAS, and provide
other necessary information (channel, kind of data,
computer program, and TLC plate identification) via
the thumbwheel switches (TWSs). The magnetic tapes
were delivered to another building in the Conoco complex for overnight processing by a large IBM computer
installation. With this system, a smooth transition
from hand-processed to computer-processed data was
achieved without disrupting the routine sample sched-
ule.

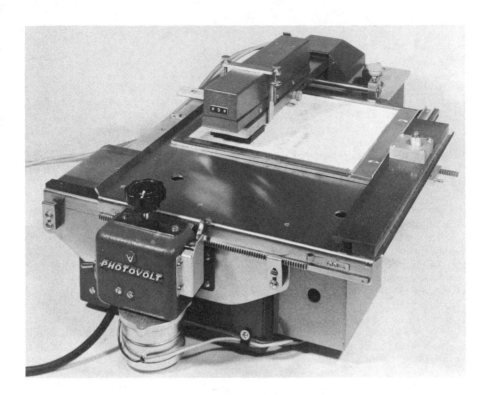

Figure 2. Automatic track-changing apparatus introduced in 1971.

AUTOMATIC TRACK CHANGER

Figure 2 shows the first instrument that was used to
automate the track-changing operation. Design and
operating details for this prototype of the present
densitometer were published in 1971 (4). Certain
scanning motor modifications and a novel clutch with
mechanical drive made this instrument capable of automatically scanning all 25 tracks and shutting itself
down at the end of the last scan.
     The most serious problem with this fully auto-
mated instrument was associated with the fact that
advancement from track to track was based on a preselected fixed increment of distance between track

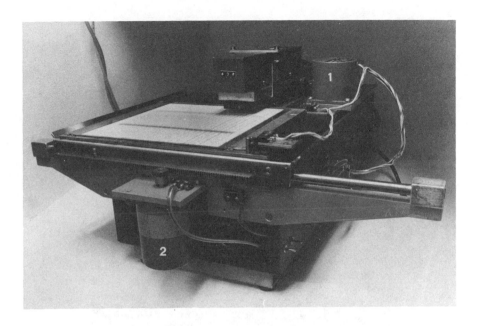

Figure 3. Present fully automatic scanning TLC densitometer installed in 1975:   (1) track-scan stepping motor
and (2) track-advance stepping motor.

centers.   At that time, it was difficult to track layers with the degree of precision required by the
instrument.   Frequent misalignment of one or more of
the last few tracks with the source slit finally
prompted other interested colleagues to see if that
problem might be resolved.   This led to the development of the present instrumentation (5) and the ulti-
mate success of the current system for data acquisition, computing, and reporting.

PRESENT DENSITOMETER

Figure 3 is a recent photograph of the present fully
automatic scanning TLC densitometer.   This particular
instrument was installed in 1975 and has already been
used to measure an estimated 250,000 track density profiles.   It is a rugged instrument and appears capable
of providing many more years of dependable service.

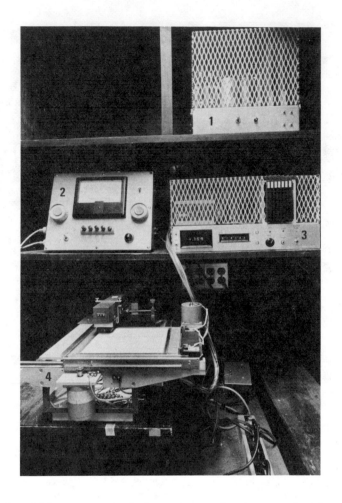

Figure 4. All current laboratory instrumentation:
(1) power supply, (2) multiplier photometer, (3) microprocessor, and (4) densitometer.

Figure 3 shows that this densitometer is equipped
with two stepping motors. When the automatic track-
changing apparatus was modified, the clutch was
replaced with Stepping Motor No. 1, and the synchronous motor was replaced with Stepping Motor No. 2.
Motor No. 1 scans each track along the y axis, and

Motor No. 2 advances the plate from track to track
along the x axis. To accomplish this, the TLC plate
and the entire optical system were rotated 90° clock-
wise in the xy plane. At the same time, the electronics for the source lamp were improved. These improvements included the use of a lower filament voltage to
increase the life of the lamp and the installation of
a photo cell for automatically monitoring and control-
ling the intensity of the incident light.

CURRENT LABORATORY INSTRUMENTATION

Figure 4 shows the densitometer and all other parts of
the current laboratory instrumentation. The entire
package is installed in a small darkroom. The power
supply is for the stepping motors. The multiplier
photometer was retained for adjusting the dark current
before each set of 25 tracks is scanned. It is also
still used to amplify the detector signal. The Varicord recorder (Figure 1) was eliminated. The old DVM
and console (Figure 1) were replaced by a digital
panel meter (DPM) and set of TWSs. These are located
on the front panel of the microprocessor.

The microprocessor was assembled from Control
Logic Model L modules utilizing the Intel 8008 central
processor unit chip. Some of the interface circuitry
was purchased from Control Logic. The rest of it was
built in-house. Programming of the microprocessor was
done in a high-level language code called PLM, a sub-
set of the more widely known PL1.

DENSITOMETER PREPARATION

Figure 5 shows only the densitometer, photometer, and
microprocessor units. A TLC plate with five charred
standard chromatograms and 20 charred sample chromatograms is on the carriage. All of the blank tracks on
this plate were scanned, and their density profiles
were recorded on magnetic tape before samples were
applied to the tracks.

The TWSs have been used (from left to right) to
dial a one-digit instrument identification code (9), a
one-digit kind of data code (2 for blank tracks and, in
this case, 1 for charred tracks), a one-digit computer

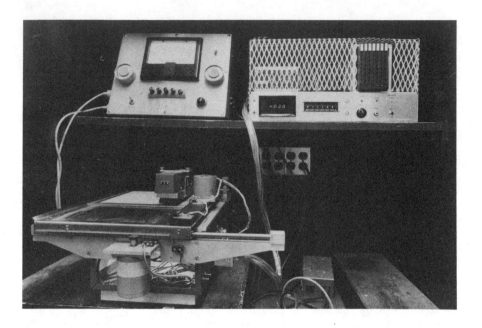

Figure 5. Densitometer, photometer, and microprocessor.

program code (1), and a three-digit TLC plate number (132). Other switches on the front panel of the microprocessor have been used to set up the densitometer for "manual" operation with "slow" stepping motor rate.

In Figure 5, toggle switches for the two stepping motors can be seen on the right-hand side (near the front) of the densitometer. These have been used to position the plate so that the beam of light is centered in, and two steps above the y stepping motor limit near the bottom of, Track 25. This is a clean zone of the layer between the final developing solvent level line and the spot point. With no light striking the detector (shutter closed), the multiplier photometer has been adjusted to give a DPM reading of zero, and with the detector exposed to transmitted light (shutter opened), the source lamp intensity has been adjusted to give a DPM reading of 1.00 ± 0.05 V.

At this point, the controls on the front panel of the microprocessor are again used. The densitometer

is switched from "manual" to "auto" scan and from
"slow" to "fast" stepping motor rate. Then the "reset"
and "go" buttons are pressed.

MICROPROCESSOR OPERATION

The microprocessor immediately picks up and sends the
TWS information to the ADAS, where it is acknowledged
and recorded on magnetic tape. The acknowledgment or
"handshake" signals the microprocessor to pick up the
first data point from the DPM and send it to the ADAS.
As soon as the "handshake" for that point is completed,
the microprocessor causes the track-scan motor to move
the plate one step in the y direction and transmit the
second DPM reading to the ADAS. This process is
repeated until 611 data points (100 points per inch of
scan) have been acquired for Track 25. These data
points include those for clean zones of the layer
which lie below the spot point and immediately below
the most advanced developing solvent front.
    The microprocessor then drives the track advance
motor to move the plate in the x direction while monitoring the output from the DPM. When the detector is
exposed to light from the source lamp passing through
the groove in the layer between the last track and the
next track to be scanned, a DPM overflow condition is
created. This initiates a step-counting process which
continues until a second flash of light, passing
through the groove on the opposite side of the next
track to be scanned, is seen. When that occurs, the x
stepping motor reverses and moves the plate one-half
as many steps in the opposite direction. By this
means, the source lamp slit and detector are precisely
aligned with the center of the next track, and this
alignment is entirely independent of small or even
very large variations in track width.
    After the track change has been completed, the
microprocessor causes the y stepping motor to take 611
steps in reverse. This is performed at high speed,
with no data being acquired.
    The program then calls for a repeat of the track
scan, track change, and fast return routines until 611
data points have been acquired from each track. When
all tracks have been scanned, the instrument automatically shuts down. The entire process takes about
8-1/4 min of unattended instrument operation.

Figure 6. Analytical data acquisition system.

ANALYTICAL DATA ACQUISITION SYSTEM (ADAS)

Figure 6 is a front view of the remote ADAS with which
the microprocessor communicates. It consists of a
Nova 1200 minicomputer; a 9-track, 1600-bit-per-inch
tape drive; a real-time clock; and an "input/output
bus expander," which allows input of binary coded deci-
mal data from six channels. The minicomputer checks

each incoming data point to determine whether it is a
valid number; flags errors which occur; tags each data
point with the channel number; blocks the data; writes
it on tape in a format suitable for the IBM installation to read; and reactivates the microprocessor program. At the end of each day, the reel of magnetic
tape is replaced and sent by an evening shift runner
to the Computer Services Division of Conoco for overnight processing.

COMPUTING

Computer Services reads the tape into a TLC data file.
The contents of this file are then sorted and indexed
by plate number. The blank track data are placed in
one file, while the charred track data are placed in
files according to the type of analysis being performed. In addition to the blank track and charred
track data, other information is required (sample numbers, standard identification and composition, etc.).
At first, this information was supplied via punched
cards which were transported by runner with the tape.
Now it is entered directly from a computer terminal
into a Computer Services disk file. Finally, appropriate combinations of special and general routines are
used to compute the results.
     Some of the computer programs are more sophisti-
cated than others, but most of them include certain
general routines. These are designed to:
     1. Smooth the raw densitometer data.
     2. Convert transmittance measurements into
        concentration-related units.
     3. Scale and subtract each blank track from
        the corresponding charred track data.
     4. Divide the net density profiles into seg-
        ments.
     5. Integrate, sum, and normalize peak areas.
     6. Calculate standard sample statistics.
     7. Correct and adjust all results to stand-
        ard sample basis.
     8. Evaluate corrected and adjusted results
        for samples and/or the standard sample.
     9. Print and plot all necessary information
        in the desired report format.

REPORTING

The Computer Services Division produces a customer as
well as a laboratory report for each plate of 25 samples. The customer report is specifically tailored to
satisfy the needs of the recipient. It contains the
requested analytical information, which can be presented in almost any form desired. A description of
the analytical method, as well as an evaluation and
interpretation of the results, also can be provided.
The laboratory report is even more comprehensive. In
addition to everything that is included in the customer report, the laboratory report contains a variety
of other information that is used for quality control.
Examples of the kinds of information in both types of
reports are shown and explained elsewhere (6, 7).
     The computer-generated reports are delivered to
the laboratory by a night shift runner. In the morning, after the reports have been checked, the customer
reports are distributed and the laboratory reports are
filed along with other pertinent documentation. The
other documentation includes TLC plate logs, lab notes,
photographs, and/or cross references to a file of
photographic negatives of the visualized chromatograms.
As additional filing space is needed, all paper laboratory records are microfilmed for long-term storage and
recall.

SUMMARY

A basic system for quantitative TLC has been under
development and in constant use by this laboratory for
the past 15 years. Current methodology for preparing
large numbers of chromatograms for this system is
described elsewhere (1). This chapter shows how the
data acquisition, computing, and reporting were automated. A microprocessor-operated TLC densitometer and
minicomputer-controlled data system are used to record
all blank track and charred track density profiles on
magnetic tape. The tape data are read and processed
by an IBM computer installation to generate the
required results and prepare the necessary reports.

ACKNOWLEDGMENTS

For their contributions to the success of the TLC
effort at Conoco Inc., we wish to thank D. C. Herrick,
A. F. Hofmann, B. Horsfield, W. P. Kane, F. Kennedy,
J. C. Kirk, D. L. Kohler, R. S. Lane, R. E. Laramy,
W. D. Leslie, D. E. Linder, D. E. Monn, H. F. Nicolay-
sen, C. E. Payton, and others.

REFERENCES

1. J. B. Sudbury, T. T. Martin, and M. C. Allen, in
   Techniques and Applications of Thin Layer Chromatography, J. C. Touchstone and J. Sherma (Eds.), Wiley,
   New York, 1984, Chapter 11.
2. T. T. Martin and M. C. Allen, J. Am. Oil Chem. Soc.
   48, 752 (1971).
3. M. C. Allen and T. T. Martin, J. Am. Oil Chem. Soc.
   48, 790 (1971).
4. G. J. Hanggi and T. T. Martin, United States Defen-
   sive Publication T893,011; Official Gazette 893,
   403, December 14, 1971.
5. D. E. Monn, R. W. Bockhorst, and H. W. Powilleit,
   "A Microprocessor Controlled Automatic Scanning
   Densitometer for Thin Layer Chromatography Plates,"
   ACS National (Fall) Meeting, San Francisco, California, August 29 through September 3, 1976.
6. T. T. Martin, M. C. Allen, and J. B. Sudbury, in
   Techniques and Applications of Thin Layer Chromatography, J. C. Touchstone and J. Sherma (Eds.), Wiley,
   New York, 1984, Chapter 13.
7. J. B. Sudbury, T. T. Martin, and M. C. Allen, in
   Techniques and Applications of Thin Layer Chromatog-
   raphy, J. C. Touchstone and J. Sherma (Eds.), Wiley,
   New York, 1984, Chapter 14.

Characterizing C_{15+} Fractions of Crude Petroleum by Thin Layer Chromatography

Ted T. Martin
Marvin C. Allen
J. Byron Sudbury

INTRODUCTION

Huc et al. (1, 2) describe an excellent, indirect-quantitative thin layer chromatography (TLC) method for compound types in small amounts of C_{15+} extracts from powdered sedimentary rocks. In that method, preparative TLC is used to separate three classes of organic material which are recovered and weighed. The recovered material can be subsequently evaluated by other analytical techniques.

In this laboratory, when recovery of the separated material is not required, a less time-consuming analytical TLC separation with in situ quantitation of the various compound types is employed. A primitive version of this direct-quantitative TLC method was published in 1973 (3). This chapter will show how a basic system for quantitative TLC (4, 5) is currently used to process large numbers of C_{15+} crude petroleum fractions in an even more cost-effective manner.

SCOPE OF METHOD

This is a direct-quantitative TLC method. It is designed to determine the distribution (relative to a

suitable standard sample) of saturated hydrocarbon
(SAT), aromatic hydrocarbon (AROM), and resin and
asphaltene (RASP) compound types in C_{15+} fractions of
crude petroleum.

The method is particularly useful when sedimentary rock extracts, extracts of some soils (6), and
other samples of crude petroleum are too small for the
satisfactory recovery of compound type fractions or
there is no need for recovery of those fractions. As
little as 0.5 mg of a well-stabilized (C_{15+}) organic
material is sufficient.

Ideally, 0.5 to 50.0 mg samples are received in
numbered 1-oz glass bottles with polyethylene-lined
screw caps and accompanied by a list of net weights.
Under these conditions, an experienced technician can
perform at least 100 to 150 analyses per day. After
overnight computing, the results are reported on the
next workday.

For a typical standard sample, the coefficient of
variation is usually less than 2 percent. The accuracy of routine sample results depends on the proper
use of an appropriate (7) and accurately analyzed or
synthesized standard sample and the number of routine
sample replicate analyses that are available for averaging.

METHODOLOGY

Most of the general methodology has already been
described (4, 5). Only the additional materials and
procedures that are used for this and certain other
methods (8) will be included here.

Standard Samples

For some applications, the speed with which routine
samples can be reproducibly characterized is of
greater importance than results that are accurate or
at least comparable to those obtained by more traditional methods. For these applications, almost any
standard sample can be selected or synthesized and
consistently used. Reduction of the usual track-to-
track and plate-to-plate errors is favored by a standard sample that contains a significant amount of each
compound type of interest. If not synthesized, the

TABLE I. SAMPLE APPLICATION CONCEPTS AND CONVENTIONS HCCl₃/MeOH (80:20 v/v) SOLNS.

A. TRACKS AND GROUPS

Track No.	1	2	3	4	5	6	7	8	9	10	11	12	13	14	15	16	17	18	19	20	21	22	23	24	25
Std Spl, 4 µℓ ≅ 12.5 µg	X						X						X						X						X
Spls, 4 µℓ = 2.5 to 12.5 µg		X	X	X	X	X		X	X	X	X	X		X	X	X	X	X		X	X	X	X	X	
Spl Group No.				1						2						3						4			

B. RULES FOR SINGLE AND REPLICATE ANALYSES

	1						2						3						4						
Spl Group No.																									
Track No.	1	2	3	4	5	6	7	8	9	10	11	12	13	14	15	16	17	18	19	20	21	22	23	24	25
Cap ID for Std Spl	0						0						0						0						0
Cap ID for 1 x 20 Spls		1	2	3	4	5		6	7	8	9	10		11	12	13	14	15		16	17	18	19	20	
Cap ID for 2 x 10 Spls		1	2	3	4	5		6	7	8	9	10		1	2	3	4	5		6	7	8	9	10	
Cap ID for 4 x 5 Spls		1	2	3	4	5		1	2	3	4	5		1	2	3	4	5		1	2	3	4	5	

standard sample can be analyzed by almost any method for compound types, including this one. Compound-type distributions for the standard and routine samples can be expressed as weight percents, but are frequently reported as simple area percents.

When the results obtained by this method must be accurate or at least compatible with those acquired by other means, the selection or synthesis of an appropriate standard sample becomes more critical. These standard samples should still contain a significant amount of each compound type, but the compositions of the compound types should be similar to those in the routine samples. Sometimes a larger than usual supply of a typical routine sample can be carefully analyzed by low-pressure column chromatography or preparative TLC and then used as a standard sample. At other times, it is more desirable to recover enough of the fractions so that weighed amounts can be recombined to obtain a standard sample that contains a more balanced concentration of each compound type.

Sample Application

Table I summarizes most of what needs to be said about application of the standard and routine samples. The solutions are prepared in chloroform-methanol (80:20 v/v). They are applied to soft silica gel G layers which have been pre-eluted, divided into 25 individual

5-mm-wide tracks, and prescanned to record the density
profile of each blank track (4, 5).   Four microliter
glass capillaries are used to load as much solute as
possible without allowing any portion of a track with
charred chromatogram to exceed an optical density of
1.3 (9).   In practice, no more than about 12.5 µg of a
standard sample or routine sample is loaded on a track.
The standard sample solution is applied to tracks 1,
7, 13, 19, and 25; and a routine sample solution is
applied to one or more of the 20 remaining tracks.
This format results in four groups of five tracks for
routine samples.
          Table I also summarizes the rules for single and
replicate analyses.   One capillary is used to load the
standard sample, and a different capillary is used for
each routine sample.   In Table I, the standard sample
capillary is identified as zero.   When a single analysis of 20 routine samples is desired, 20 capillaries
are used to load these samples as indicated.   Capillaries 1-5, 6-10, 11-15, and 16-20 are used to load
them on the tracks in groups 1 through 4, respectively.
When duplicate analyses for 10 routine samples are
desired, 10 capillaries are used to load the samples
on tracks in the first two and the last two groups as
shown.   When quadruplicate analyses for five routine
samples are desired, only five capillaries are
required.   Each of the five samples is loaded on a corresponding track in each of the four groups.

          Developing

Two different solvent mixtures are sequentially used
to develop the chromatograms.   For the first development, chloroform-methanol-acetic acid (80:20:1) is
allowed to migrate until the solvent front is 65 mm
above the bottom edge of the plate.   This moves most
of the sample material away from the application
points.   For the second development, petroleum ether-
toluene (100:4) is allowed to migrate until the solvent front is 171 mm beyond the bottom edge of the
plate.   This moves the hydrocarbons (SAT + AROM) away
from the so-called nonhydrocarbons (RASP) and separates the SAT from the AROM compound types.
          Each development is performed in a filter paper
lined tank after 20 min of layer-solvent vapor equilibration.   Between developments, the layer is dried for

10 min in a 120° C vacuum oven maintained at 400 mbar
(12 in. of Hg) with a filtered air sweep and allowed
to cool for 10 min in a desiccator. After the final
development, the layer is dried for 15 min in a 130° C
vacuum oven with the same level of reduced pressure
and air sweep. At the end of that step, the plate is
immediately transferred to the char block.

Subsequent Steps

The chromatograms are made visible by charring. This
is accomplished by heating the plate for 20 min at
125° C in an atmosphere of SO_3 as described earlier
(4). Data acquisition, computing, and reporting are
also carried out as previously described (5).

EXAMPLE

An example is used to illustrate the method. For this
example, traditional methods were employed to separate
a sample of C_{15+} crude petroleum into the compound
types of interest. The recovered fractions were used
to prepare a mixture containing 40, 44, and 16 weight
percent of the SAT, AROM, and RASP compound types,
respectively. Six identical spotting solutions of the
mixture were then prepared so that each would contain
2.5 µg/µl. One of these solutions was selected to
represent a standard sample. The others were treated
as if they were spotting solutions of five different
routine samples. The results are presented and discussed.

The Chromatograms

Figure 1 shows the 25 charred chromatograms and identifies the SAT, AROM, and RASP compound-type zones. It
also indicates where each sample was applied and summarizes some of the developing details. In this figure, two of the arrows are intended to emphasize the
importance of using an initial solvent level (ISL) for
the second development that is above the ISL of the
first (4). Failure to observe this rule can result in
the contamination of chromatograms with a variety of
impurities which are adsorbed from the first solvent,
eluted by the second, and ultimately charred. The

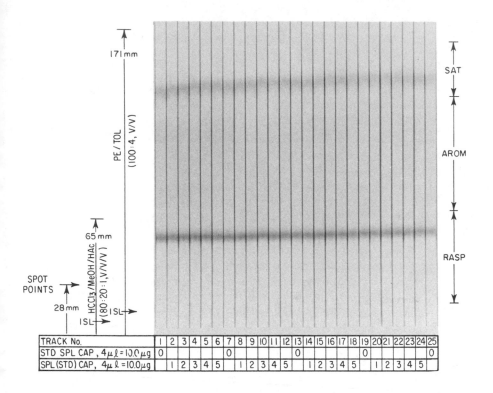

Figure 1. The 25 charred chromatograms.

effect of this kind of contamination cannot be removed
by subtracting the density profiles of blank tracks
before sample application from the density profiles of
those tracks after char (5).

At this stage, the laboratory work was essentially complete. Even before the photograph for Figure
1 was taken, all of the charred track data, as well as
the blank track data, had been automatically acquired
and sent with the other necessary information for IBM
computing and reporting (5). The remainder of this
discussion will deal with various parts of the computer-generated laboratory report.

Example   181

TABLE II. SOME OF THE MORE BASIC RAW DATA

PLATE NO. 3306

1	40	160	396	590	611	15.2999	24.7556	18.9788
2	40	158	395	576	611	15.3901	24.9743	19.1023
3	24	157	395	573	611	16.3063	26.3171	19.6397
4	40	156	395	573	611	16.6556	26.6404	19.7052
5	25	155	395	574	611	17.1334	28.3157	20.8485
6	40	154	395	576	611	16.9571	28.2837	21.0053
7	40	153	395	583	611	17.5867	29.4219	21.2548
8	34	151	394	575	611	18.4529	31.0731	22.1122
9	38	150	394	577	611	18.1012	30.8847	21.7383
10	3	149	393	577	611	19.6532	32.6418	21.9228
11	22	148	393	581	611	18.2492	30.5000	22.2789
12	38	147	392	576	611	18.3568	31.2752	22.0186
13	12	146	392	577	611	18.7713	31.9003	22.7868
14	17	144	391	577	611	18.3627	31.3236	22.9260
15	28	142	390	576	611	17.6708	30.0056	22.4447
16	3	140	390	572	611	17.7247	29.7176	22.9449
17	16	138	389	576	611	16.9330	30.0294	23.5250
18	23	136	388	573	611	16.3433	28.2849	22.9095
19	18	135	388	577	611	15.9229	27.5790	22.4998
20	21	135	387	579	611	15.6917	25.9707	22.0096
21	21	135	387	577	611	15.5809	27.1727	22.1268
22	19	135	386	576	611	15.1520	26.3840	21.8928
23	22	135	386	577	611	15.2150	25.9359	21.0120
24	17	135	385	576	611	15.3472	25.9825	21.7881
25	32	135	385	576	611	14.1898	25.7782	21.8106

Quality Control

Because of the relative ease with which various kinds
of results can be made available, much more quality
control information is obtained than can be readily
described here. Only a small portion of it will be
discussed. Some of the more basic raw data are presented in Table II.

Table II shows the one-digit computer program
code followed by the three-digit TLC plate number and
nine columns of numerical information. The first of
the nine columns lists the 25 track numbers. Column 6
merely confirms the fact that 611 data points were
read from each track.

Columns 2 and 5 show the least dense points in
the smoothed data for the short sections of each blank
track that were above and below the portion reserved
for the chromatogram. They are the so-called scaling
points. The data for those points were used by the
computer to position the blank track baselines so that
smoothed net density profiles for the charred chromatograms could be obtained. The scaling points are also
used to define the beginning and end of the chromatograms.

C_{15+} Fractions of Crude Petroleum

TABLE III. SAMPLE OF DATA
FOR PARTIALLY CORRECTED RESULTS

STD 1W-0	3306	1	25.92	41.93	32.15	59.03
4936-070-001	3306	2	25.88	42.00	32.12	59.47
4936-070-002	3306	3	26.19	42.27	31.54	62.26
4936-070-003	3306	4	26.44	42.29	31.28	63.00
4936-070-004	3306	5	25.84	42.71	31.45	66.30
4936-070-005	3306	6	25.60	42.69	31.71	66.25
STD 1W-0	3306	7	25.76	43.10	31.14	68.26
4936-070-001	3306	8	25.76	43.38	30.87	71.64
4936-070-002	3306	9	25.59	43.67	30.74	70.72
4936-070-003	3306	10	26.48	43.98	29.54	74.22
4936-070-004	3306	11	25.69	42.94	31.37	71.03
4936-070-005	3306	12	25.62	43.65	30.73	71.65
STD 1W-0	3306	13	25.55	43.43	31.02	73.46
4936-070-001	3306	14	25.29	43.14	31.57	72.61
4936-070-002	3306	15	25.20	42.79	32.01	70.12
4936-070-003	3306	16	25.18	42.22	32.60	70.39
4936-070-004	3306	17	24.02	42.60	33.37	70.49
4936-070-005	3306	18	24.20	41.88	33.92	67.54
STD 1W-0	3306	19	24.13	41.79	34.09	66.00
4936-070-001	3306	20	24.64	40.79	34.57	63.67
4936-070-002	3306	21	24.01	41.88	34.10	64.88
4936-070-003	3306	22	23.89	41.60	34.52	63.43
4936-070-004	3306	23	24.48	41.72	33.80	62.16
4936-070-005	3306	24	24.32	41.17	34.52	63.12
STD 1W-0	3306	25	22.97	41.73	35.30	61.78

     In columns 3 and 4, the data points for each
track mark the beginning and end of the AROM compound-
type zones. For the five standard sample chromatograms (tracks 1, 7, 13, 19, and 25), the computer
looks for the SAT and RASP peaks and then searches the
AROM side of the data for those peaks until the abrupt
changes in slopes are found. Extrapolation between
adjacent standard sample chromatograms is used to
establish the location of data points which signal the
beginning and end of the AROM compound-type zones for
the routine sample chromatograms.
     Columns 7, 8, and 9 list the net peak areas
obtained for the SAT, AROM, and RASP compound types,
respectively. These peak areas are expressed in
arbitrary units which are based on density measurements for the data points in the compound-type zones
of the charred chromatograms. The absolute values of
these peak areas contain a variety of systemic (plate-
to-plate and track-to-track) errors. Some of these
errors (sample load and track width variation, and the
horizontal component of a plate char-temperature gradient) can be easily removed by normalization of the

Example   183

data so that the three peak areas for each chromatogram are expressed as a percentage of the total peak
area (TPA) for that chromatogram (10).

   Table III displays the normalized peak areas
before they are corrected for carbon yield and most of
the remaining systemic errors. Two columns will be
recognized as containing the TLC plate and track iden-
tifications. The alphanumeric information on the left
side of those two columns identifies the material
applied to each track. The first three columns on the
right-hand side of the track numbers show (from left
to right) the net peak area percentages for the SAT,
AROM, and RASP compound types, respectively. The TPA
values for each track are tabulated in the last column.

   The information in Tables II and III is used only
by the TLC laboratory for quality control. Before the
analyses are reported, a separate computer routine
removes most of the remaining systemic errors (absolute char temperature variations and the vertical component of a plate char-temperature gradient) from the
results in the last four columns of Table III. At the
same time, the results are adjusted to an appropriate
or customer-approved standard sample basis. For this
example, the standard had been prepared to contain 40,
44, and 16 weight percent of the SAT, AROM, and RASP
compound types, respectively. Without overload, a
reasonable maximum TPA for the standard sample was
estimated to be about 85 arbitrary units per track.

   The computer treats the results in Table III as
if there were four separate groups of seven chromatograms (tracks 1-7, 7-13, 13-19, and 19-25). Each of
those groups contains five routine sample chromatograms which are sandwiched between two standard sample
chromatograms. The specified values for the standard
sample compound types and TPA are then simply divided
by the corresponding values obtained by averaging the
Table III results for each pair of standard sample
chromatograms. This procedure provides a set of 16
data conversion (multiplication) factors for correcting the Table III results and adjusting them to the
standard sample basis.

   Report Sheets

For this example, it was requested that the laboratory
report include a sample report sheet for each of the

TABLE IV. REPORT SHEET FOR SINGLE ANALYSES

TLRETLC PLATE NO. 3306   11/11/82

DATA CONVERSION FACTORS

GROUP 1 TRACKS	1- 7	1.5480	1.0349	0.5056	1.3355
GROUP 2 TRACKS	7-13	1.5589	1.0170	0.5148	1.1995
GROUP 3 TRACKS	13-19	1.6104	1.0327	0.4915	1.2190
GROUP 4 TRACKS	19-25	1.6987	1.0537	0.4611	1.3304

STANDARD STATISTICS (N=8)

ARTH. MEAN	39.997	43.999	16.004	85.000
STD. DEV.	0.528	0.364	0.574	4.669
REL. SD. (%)	1.319	0.827	3.587	5.493

* GEOM. MEAN = 1.6% *

SAMPLE	PLATE	TK	SAT	AROM	RASP	TPA
STD 1W-0	3306	1	40.211	43.496	16.293	78.837
4936-070-001	3306	2	40.156	43.563	16.281	79.415
4936-070-002	3306	3	40.447	43.640	15.913	83.149
4936-070-003	3306	4	40.721	43.543	15.737	84.135
4936-070-004	3306	5	39.963	44.153	15.884	88.537
4936-070-005	3306	6	39.687	44.254	16.059	88.468
STD 1W-0	3306	7	39.790	44.502	15.708	91.163
STD 1W-0	3306	7	40.152	43.822	16.026	81.884
4936-070-001	3306	8	40.092	44.043	15.866	85.932
4936-070-002	3306	9	39.845	44.352	15.803	84.836
4936-070-003	3306	10	40.785	44.191	15.024	89.027
4936-070-004	3306	11	40.104	43.727	16.169	85.201
4936-070-005	3306	12	39.879	44.325	15.797	85.947
STD 1W-0	3306	13	39.848	44.178	15.974	88.116
STD 1W-0	3306	13	40.645	44.297	15.058	89.545
4936-070-001	3306	14	40.404	44.200	15.396	88.513
4936-070-002	3306	15	40.378	43.970	15.653	85.477
4936-070-003	3306	16	40.481	43.526	15.993	85.801
4936-070-004	3306	17	39.042	44.403	16.554	85.923
4936-070-005	3306	18	39.406	43.736	16.858	82.328
STD 1W-0	3306	19	39.339	43.696	16.965	80.455
STD 1W-0	3306	19	40.684	43.711	15.606	87.809
4936-070-001	3306	20	41.539	42.645	15.816	84.710
4936-070-002	3306	21	40.530	43.845	15.624	86.317
4936-070-003	3306	22	40.447	43.688	15.864	84.386
4936-070-004	3306	23	41.114	43.473	15.413	82.702
4936-070-005	3306	24	41.058	43.118	15.823	83.972
STD 1W-0	3306	25	39.306	44.294	16.400	82.191

three Table I spotting options. Ordinarily, only one
report sheet would have been printed and sent to the
customer. In this case, it would have been the one
for five samples analyzed in quadruplicate. To illustrate the different formats, all three of the sample
report sheets are included here.

Example   185

Table IV is the report sheet for 20 samples which have been analyzed only one time. The report is identified by analysis code (TLRETLC), TLC plate number, and date the results were computed. Again, the actual three-digit plate number is preceded by the one-digit computer program code.

More than the others, the Table IV report format tends to preserve the "4 group x 7 chromatogram concept" with all the rows and columns identified. The 16 data conversion (multiplication) factors for each group and set of tracks are tabulated. The statistics are for the eight corrected and adjusted rather than the five actual standard sample analyses.

Because of the way the computer routine is written, the mean values for the standard sample are the same (within the limits of normal-precision computing) as those specified for the standard sample; and all of the routine sample analyses are similarly corrected and adjusted to standard sample basis. As a single indicator of the overall accuracy and/or precision of the routine sample results, the geometric mean of the coefficients of variation for the three compound types in the standard sample has proved most useful.

Double-precision computing cannot be justified for any of the routine samples that are analyzed by this method and was not used for this example. On the other hand, at least two insignificant figures are always provided. This is done to accommodate those who wish to submit their own standard samples as routine samples and work up their own statistics.

Table V is the report sheet for 10 samples that have been analyzed in duplicate. It does not list the data conversion factors, but does contain all of the other Table IV information. In Table V, the eight corrected and adjusted standard sample analyses, as well as their statistics, are found near the top of the report sheet. On the lower left side of the report sheet, appropriate pairs of the same standard sample analyses are also listed as duplicates along with the duplicate analyses for the 10 routine samples. The average result for each pair of duplicate analyses is shown on the lower right-hand side of the report sheet.

Table VI shows the proper report sheet for the Table I spotting option used to illustrate this method. This sheet is designed to report the results obtained when five different routine samples are analyzed in

TABLE V. REPORT SHEET FOR DUPLICATE ANALYSES

11/11/82

TLRETLC PLATE NO. 3306

SAMPLE	TK	SAT	AROM	RASP	TPA
STD 1W-0	1	40.211	43.496	16.293	78.837
STD 1W-0	7	39.790	44.502	15.708	91.163
STD 1W-0	7	40.152	43.822	16.026	81.884
STD 1W-0	13	40.845	44.178	15.974	89.116
STD 1W-0	13	40.845	44.297	15.058	89.545
STD 1W-0	19	39.339	43.696	15.965	80.455
STD 1W-0	19	39.684	43.711	15.606	87.809
STD 1W-0	25	39.306	44.294	16.400	82.191

STANDARD STATISTICS(N=8)

	SAT	AROM	RASP	TPA
ARITH MEAN	39.997	43.999	16.004	85.000
STD DEV	0.528	0.364	0.574	4.669
REL. SD. (%)	1.318	0.827	3.587	5.493
* GEOM. MEAN = 1.6% *				

SAMPLE	TK	SAT	AROM	RASP	TPA	AVG SAT	AVG AROM	AVG RASP	AVG TPA
STD 1W-0	1	40.211	43.496	16.293	78.837	40.428	43.896	15.676	84.191
STD 1W-0	13	40.645	44.297	15.058	89.545				
4936-070-001	2	40.156	43.563	16.281	79.415	40.280	43.882	15.838	83.964
4936-070-001	14	40.404	44.200	15.396	88.513				
4936-070-002	3	40.447	43.640	15.913	83.149	40.412	43.805	15.783	84.313
4936-070-002	15	40.378	43.970	15.653	85.477				
4936-070-003	4	40.721	43.543	15.737	84.135	40.601	43.534	15.865	84.968
4936-070-003	16	40.481	43.526	15.993	85.801				
4936-070-004	5	39.963	44.153	15.884	88.537	39.503	44.278	16.219	87.230
4936-070-004	17	39.042	44.403	16.554	85.923				
4936-070-005	6	39.687	44.254	16.059	88.468	39.546	43.995	16.459	85.398
4936-070-005	18	39.406	43.736	16.858	82.328				
STD 1W-0	7	39.790	44.502	15.708	91.163	39.564	44.099	16.337	85.809
STD 1W-0	19	39.339	43.696	16.965	80.455				
STD 1W-0	7	40.152	43.822	16.026	81.884	40.418	43.766	15.816	84.847
STD 1W-0	19	40.684	43.711	15.606	87.809				
4936-070-001	8	40.092	44.043	15.866	85.932	40.815	43.344	15.841	85.321
4936-070-001	20	41.539	42.645	15.816	84.710				
4936-070-002	9	39.845	44.352	15.803	84.836	40.188	44.099	15.714	85.577
4936-070-002	21	40.530	43.845	15.624	86.317				
4936-070-003	10	40.785	44.191	15.024	89.027	40.616	43.940	15.444	86.706
4936-070-003	22	40.447	43.688	15.864	84.386				
4936-070-004	11	40.104	43.727	16.169	85.201	40.609	43.600	15.791	83.951
4936-070-004	23	41.114	43.473	15.413	82.702				
4936-070-005	12	39.879	44.325	15.797	85.947	40.469	43.721	15.810	84.960
4936-070-005	24	41.058	43.118	15.823	83.972				
STD 1W-0	13	39.848	44.178	15.974	88.116	39.577	44.236	16.187	85.153
STD 1W-0	25	39.306	44.294	16.400	82.191				

186

Example   187

quadruplicate. The general format is similar to the
one used for duplicate analyses (Table V), but appropriate sets of quadruplicate rather than duplicate
analyses are listed and averaged. In addition, this
report sheet includes as much statistical information
for the sets of quadruplicate analyses as it does for
the eight corrected and adjusted standard sample analyses.

For a variety of reasons, all three of the report
sheets include TPA information. The TPA statistics
for the standard sample are used as a general check on
spotting precision. If those values seem reasonable
and equal loads of the standard sample and a routine
sample have been spotted, a gross difference between
the TPA values signals a possible problem. When no
analytical problem can be found, the original net
weight of the routine sample is checked. If the sample has been properly stabilized and the weight is correct, a low TPA for a rock extract usually indicates
that deceptively high concentrations of elemental sulfur, inorganic salts, and/or other inorganic materials
are present.

Computer Plots

When accuracy is required and reliable reference
standards are being used, automatic data handling and
computing virtually eliminate the quantitative need
for densitometer traces. However, the traditional
qualitative importance of such traces remains undimin-
ished. To be reasonably sure that the analyses are
accurate as well as precise and as a general check on
the integrity of routine samples, densitometer traces
of the chromatograms on tracks 1 through 24 of each
plate are routinely plotted by the computer, included
in the report, and carefully inspected.

Obvious qualitative differences between the computer plots for the reference standard and those for
the routine samples may indicate that a more representative reference standard is required (7). They also
may indicate that the routine samples have been mis-
labeled or are contaminated. Serious contamination of
rock extracts with drilling mud additives is possible.
In most instances, mud additives are easily detected
and recognized in the RASP compound-type region of the
chromatograms.

TABLE VI. REPORT SHEET FOR QUADRUPLICATE ANALYSES

TLRETLC PLATE NO. 3306

11/11/82

Distribution Data

SAMPLE	GROUP	TRACK	DISTRIBUTION, WT% SAT	AROM	RASP	DIVISIONS TPA
STD 1W-0	1	1	40.211	43.496	16.293	78.837
STD 1W-0	2	7	39.790	44.502	15.708	91.163
STD 1W-0	3	13	40.152	43.822	16.026	81.884
STD 1W-0	4	19	39.845	44.178	15.974	88.116
STD 1W-0	1	1	40.211	43.496	16.293	78.837
STD 1W-0	2	7	40.152	43.822	16.026	81.884
STD 1W-0	3	13	40.645	44.297	15.058	89.545
STD 1W-0	4	19	40.684	43.711	15.606	87.809
4936-070-001	1	2	40.156	43.563	16.281	79.415
4936-070-001	2	8	40.092	44.043	15.866	85.932
4936-070-001	3	14	40.440	44.200	15.396	88.513
4936-070-001	4	20	41.539	42.645	15.816	84.710
4936-070-002	1	3	40.447	43.640	15.913	83.149
4936-070-002	2	9	39.845	44.352	15.803	84.836
4936-070-002	3	15	40.378	43.970	15.853	85.477
4936-070-002	4	21	40.550	43.845	15.624	86.317
4936-070-003	1	4	40.721	43.543	15.737	84.135
4936-070-003	2	10	40.785	44.191	15.024	89.027
4936-070-003	3	16	40.481	43.526	15.993	84.801
4936-070-003	4	22	40.447	43.688	15.864	84.386
4936-070-004	1	5	39.963	44.153	15.884	88.537
4936-070-004	2	11	40.104	43.727	16.169	85.201
4936-070-004	3	17	40.042	44.403	16.554	85.923
4936-070-004	4	23	41.114	43.473	15.413	82.702
4936-070-005	1	6	39.687	44.254	16.059	85.468
4936-070-005	2	12	39.879	44.325	15.797	88.947
4936-070-005	3	18	39.406	43.736	16.858	82.328
4936-070-005	4	24	41.058	43.118	15.823	83.972
STD 1W-0	1	7	39.790	44.502	15.708	91.163
STD 1W-0	2	13	39.848	44.178	15.974	88.116
STD 1W-0	3	19	39.339	43.696	16.965	80.455
STD 1W-0	4	25	39.306	44.294	16.400	82.191

Statistics

SAMPLE	PARAMETER	SAT	AROM	RASP	TPA
STD 1W-0	X BAR	39.997	43.999	16.004	85.000
	SD	0.528	0.364	0.574	4.669
	CV	1.319	0.827	3.587	5.493
	* GEOM. MEAN = 1.6% *				
STD 1W-0	X BAR	40.423	43.831	15.746	84.519
	SD	0.280	0.338	0.538	5.010
	CV	0.692	0.772	3.419	5.928
	* GEOM. MEAN = 1.2% *				
4936-070-001	X BAR	40.548	43.613	15.839	84.643
	SD	0.674	0.699	0.362	3.829
	CV	1.663	1.604	2.284	4.523
	* GEOM. MEAN = 1.8% *				
4936-070-002	X BAR	40.300	43.952	15.748	84.945
	SD	0.309	0.299	0.135	1.341
	CV	0.767	0.680	0.856	1.579
	* GEOM. MEAN = 0.8% *				
4936-070-003	X BAR	40.608	43.737	15.655	85.837
	SD	0.169	0.311	0.433	2.249
	CV	0.415	0.712	2.767	2.620
	* GEOM. MEAN = 0.9% *				
4936-070-004	X BAR	40.056	43.939	16.005	85.591
	SD	0.848	0.417	0.481	2.400
	CV	2.117	0.950	3.004	2.804
	* GEOM. MEAN = 1.8% *				
4936-070-005	X BAR	40.008	43.858	16.134	85.179
	SD	0.727	0.559	0.497	2.645
	CV	1.817	1.274	3.080	3.105
	* GEOM. MEAN = 1.9% *				
STD 1W-0	X BAR	39.571	44.167	16.262	85.481
	SD	0.287	0.341	0.549	5.010
	CV	0.726	0.773	3.374	5.861
	* GEOM. MEAN = 1.2% *				

Example   189

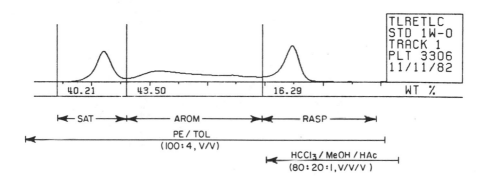

Figure 2. The computer plot.

Figure 2 features only one of the computer plots. It shows the blank-track-background-corrected net density profile for the standard sample chromatogram on track 1 of the TLC plate in Figure 1.

In Figure 2, the composition of the first and second developing solvents is given. Also, the direction and distance of development with each of the developing solvents is indicated. The baseline scale marks the distance between each 50 of the 611 data points. The data point numbers increase from left to right.

The computer plot in Figure 2 is divided into three segments at four baseline locations. These locations are identified in Table II by data point numbers. For the Figure 2 plot, the Track 1 data point numbers are found in columns 2 through 5 of Table II to be 40, 160, 396, and 590. In Figure 2, a vertical line is used to mark the location of only the first three of these four data points. From left to right, the three segments represent the SAT, AROM, and RASP compound types. The application point is located at the right-hand side of the RASP peak, and materials remaining there are included with the RASP compound types.

Computer integration of the data plotted for the SAT, AROM, and RASP compound types in Figure 2 was used to obtain the track 1 peak areas that are listed in the last three columns of Table II. On the other hand, the compound-type weight percent values in

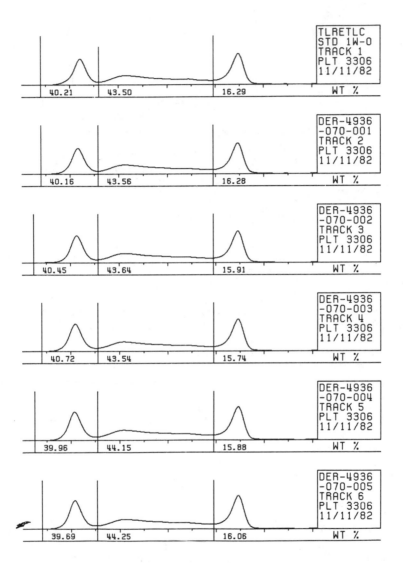

Figure 3. Computer plots for group 1 (tracks 1-6) chromatograms.

Example   191

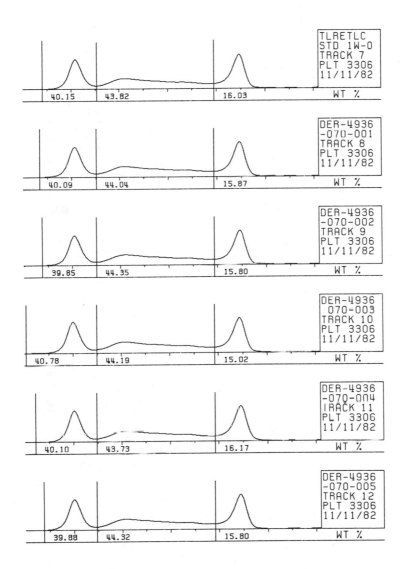

Figure 4. Computer plots for group 2 (tracks 7-12) chromatograms.

C_{15+} Fractions of Crude Petroleum

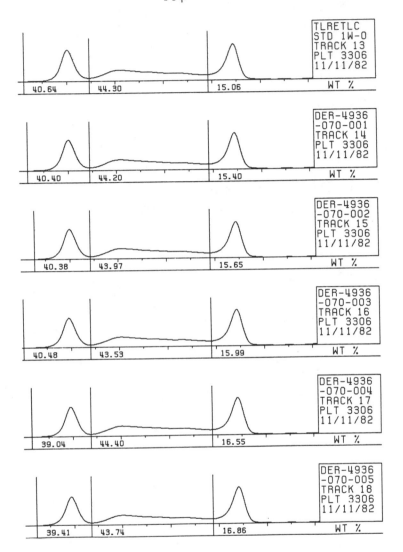

Figure 5. Computer plots for group 3 (tracks 13-18) chromatograms.

Example   193

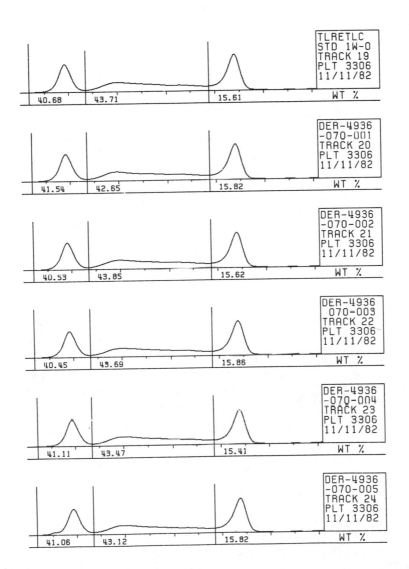

Figure 6. Computer plots for group 4 (tracks 19-24)
chromatograms.

Figure 2 have been corrected for systemic errors and
adjusted to standard sample basis. They are the same
as those listed for Track 1 in Tables IV, V, and VI.
    In Figure 2, the left edge of the block at the
right-hand end of the plot coincides with the 611th
data point. This block contains the analytical method
code and identifies the sample, track, plate, and
computer date.
    Figures 3 through 6 show the four 8-1/2 x 11 in.
pages of computer plots from the laboratory report.
Each successive page contains plots for the chromatograms on the first six tracks in each successive group
as is most clearly indicated in Table IV. Since the
SAT, AROM, and RASP weight percent values for the
track 1 plot are taken from group 1, the values for
the tracks 7, 13, and 19 plots are taken from groups
2, 3, and 4, respectively. The computer plot for the
track 25 standard sample chromatogram is omitted.

SUMMARY

This chapter shows how a basic system for quantitative
TLC (4, 5) is used by an experienced technician to
obtain at least 100 to 150 analyses per day while
determining the distribution of compound types (saturated hydrocarbons, aromatic hydrocarbons, and resins
and asphaltenes) in a variety of small (0.5 to 50.0
mg), well-stabilized (C_{15+}) samples of crude petroleum.
An example is used to illustrate the method. Various
parts of the computer-generated laboratory report are
presented and discussed.

ACKNOWLEDGMENTS

For their contributions to the success of the TLC
effort at Conoco Inc., we wish to thank G. Perkins Jr.,
C. D. Pitts, H. W. Powilleit, E. G. Poynor, O. S.
Privett, W. C. Pusey III, P. S. Randall, D. E. Robison,
A. S. Rosenberg, E. H. Schmauch, P. A. Schwab, W. L.
Sievers, C. C. Smith, J. B. Smith, and others.

REFERENCES

1. A. Y. Huc, J. Roucaché, M. Bernon, G. Caillet,
   M. da Silva, Rev., Inst. Fr. Pet. 31, 67 (1976).
2. A. Y. Huc and J. G. Roucaché, Anal. Chem. 53, 914
   (1981).
3. J. B. Smith and T. T. Martin, British Patent
   1,330,804, September 19, 1973.
4. J. B. Sudbury, T. T. Martin, and M. C. Allen, in
   Techniques and Applications of Thin Layer Chromatography, J. C. Touchstone and J. Sherma (Eds.),
   Wiley, New York, 1984, Chapter 11.
5. T. T. Martin, M. C. Allen, and J. B. Sudbury, in
   Techniques and Applications of Thin Layer Chromatography, J. C. Touchstone and J. Sherma (Eds.),
   Wiley, New York, 1984, Chapter 12.
6. J. D. Meyers and R. L. Huddleston, 34th Purdue Ind.
   Waste Conf., J.M. Bell (Ed.), Ann Arbor Science, Ann
   Arbor, Michigan, 1980, p. 686.
7. L. Hunter, Environ. Sci. Technol. 9, 241 (1975).
8. J. B. Sudbury, T. T. Martin, and M. C. Allen, in
   Techniques and Applications of Thin Layer Chromatography, J. C. Touchstone and J. Sherma (Eds.),
   Wiley, New York, 1984, Chapter 14.
9. D. T. Downing, J. Chromatog. 38, 91 (1968).
10. M. C. Allen and T. T. Martin, J. Am. Oil Chem. Soc.
    48, 790 (1971).

Characterizing High-Boiling Fractions of Refined Petroleum by Thin Layer Chromatography

J. Byron Sudbury
Ted T. Martin
Marvin C. Allen

INTRODUCTION

Amos (1) discussed the utility of thin layer chromatography (TLC) in analyzing heavy distillates, residues,
and crudes, but primarily seemed to consider TLC of
petroleum samples as "a valuable screening technique
for HPLC analysis." Wernicke et al. (2) described a
rapid TLC system for structural group analysis of
heavy hydrocarbon fractions; their TLC values compared
favorably with results obtained by column chromatography and by high-pressure liquid chromatography
(HPLC). Ray et al. (3) described a TLC procedure for
the semiquantitative analysis of compound types present in petroleum products with boiling points above
300° C. Their techniques are interesting, and their
discussion of problems encountered while analyzing
heavy refined petroleum samples is certainly pertinent
to this chapter. A similar approach has been applied
to the analysis of coal-derived liquids (3, 4).

A basic TLC system for preparing, visualizing,
and densitometering chromatograms has been described
along with our general computer data handling procedures (5, 6). This basic TLC system can be used in
quantitative or semiquantitative applications, depending on the appropriateness of the standard selected

197

for the samples being analyzed. Using this system, a
description of a quantitative TLC method for compound-
type characterization of source rocks was presented
(7); by using a standard representative of the samples,
that method demonstrated the TLC system's potential
for accuracy as well as for precision. In this current chapter on characterizing coker feedstocks and
other heavy refined petroleum fractions, we have
elected to use the same basic TLC system in a semiquantitative manner. Use of a single standard to
evaluate a variety of sample types limits the absolute
accuracy of the results, but consistent use of the
same standard permits results obtained for similar samples to be favorably compared.

   SCOPE OF METHOD

This direct-semiquantitative TLC method is used to
characterize various high-boiling fractions of refined
petroleum relative to a suitable standard sample.
This method has been applied to many sample types
including the following: certain crude oils, asphalt,
pitch, ASCO bottoms, coal tar, thermal tar, flashed
thermal tar, pyrolysis tar, coker oil, gas oil, heavy
gas oil, hydrotreated gas oil, vacuum resids, and fuel
oils. The method has also been used to characterize
samples of coal-derived liquids and various other high-
boiling mixtures.
   The method results in zones on the chromatogram
containing the following compound types: nonaromatic
hydrocarbons (alkanes and alkenes), aromatic hydrocarbons, polar compounds (resins, most of the NSO species, and some pentane-insoluble asphaltenes), and
spot point materials (pentane-insoluble asphaltenes,
and other extremely polar and/or relatively insoluble
materials). When they are present, this method also
results in a separate zone between the nonaromatic and
aromatic hydrocarbon zones containing certain naphthenes, olefins, and/or alkylbenzenes.
   Sample results in area percent are obtained relative to the known standard sample used. The standard
sample can be one whose composition has been determined by other techniques, or can be synthesized from
available compound type fractions. The precision of
the sample results is determined by the number of

sample replicate analyses performed. For a typical
standard sample (n = 8), the geometric mean of the
coefficients of variation is usually less than 3 percent.

EXPERIMENTAL

Apparatus

The basic equipment used for TLC chromatograph preparation and visualization (5) and densitometry (6) has
previously been described in detail. The TLC separations were performed on 250-μm silica gel G 20 x 20 cm
Classic UNIPLATES™ (Analtech).

Reagents

Chloroform (with preservative), ethyl ether (with preservative), methanol, and petroleum ether (30-60° C)
were all distilled in glass (Burdick and Jackson).
Glacial acetic acid (Reagent ACS) and fuming sulfuric
acid (Reagent, 30 percent SO_3) were obtained from
Fisher Scientific.

Standard Sample

When this method is used to evaluate sample compositions relative to each other in a given sample set, or
to obtain sample "fingerprints," one standard sample
will suffice to evaluate the whole gamut of sample
types available. Although absolute accuracy is not
achieved by such semiquantitative use of this TLC
system, the relative results are rapidly obtained,
reproducible, and sufficient for many practical applications. By using the same standard consistently, the
results obtained over time for samples of a given type
can be successfully related to each other.

Sample and Standard Preparation and Application

A sample size of 0.040 g is weighed into a tared 1-oz
bottle with a polyseal lid. Liquid samples may be
transferred by pipet whereas the solid samples have to
be chipped or crushed and transferred with a spatula.
The samples are dissolved in 10 ml of chloroform-

methanol (8:2, v/v). The standard is also prepared to
the same concentration (4 µg/µl) in this solvent.
     The standard is applied with a 4-µl capillary to
track numbers 1, 7, 13, 19, and 25 on a pre-eluted,
tracked, and prescanned plate prepared as described
earlier (5). A different 4-µl capillary is used to
transfer each sample to the appropriate track(s), the
tracks being determined by whether single, duplicate,
or quadruplicate analyses are desired as previously
described (7).

Chromatogram Development and Visualization

The chromatograms are prepared using three successive
developments. After standard and sample application,
the TLC plate is placed layer side down in a TLC
developing tank containing petroleum ether-chloroform-methanol
acetic acid(200:80:20:1, v/v/v/v) 12 mm deep.   Initially, the tank
is tilted to permit the layer to equilibrate with the
solvent vapor. After a 20-min equilibration period,
the tank is set upright and the solvent permitted to
migrate up the layer 39 mm past the sample application
point. The plate is then removed, and the layer is
permitted to dry in the hood for 10 min. The plate is
then placed in a tank with 14 mm of petroleum ether
and the layer permitted to equilibrate for 20 min
prior to developing until the solvent migrates 105 mm
above the sample application point. The plate is
again removed, and the layer is dried in a hood for 10
min. The layer is then equilibrated in a tank with
petroleum ether (15 mm) for 20 min prior to developing
until the solvent migrates 150 mm above the sample
application point.
     Following any one of these development or drying
steps, the chromatogra   can be visualized by exposure to short (254 nm) and/or long (366 nm) wavelength
ultraviolet light. The predominant aromatic character
of the samples makes them very responsive to this visualization technique. A color photograph taken while
inducing these chromatograms to fluoresce can reveal a
great deal about the samples, and provides an excellent qualitative record of the chromatograms.
     After the final development, the layer is placed
in an air-swept vacuum oven at 130° C to dry for an
hour. An hour drying time is adequate for our current
variety of samples. If all sample components are

extremely heavy, a drying time as short as 15 min may
be sufficient. (Inadequate drying leaves some vola-
tile sample components on the layer. During fuming
SO_3 char, these components are vaporized, redeposited
on the layer, and charred. This results in a grey-
tinted layer background. Upon densitometry, this
layer discoloration yields an elevated, "noisy"
chromatogram baseline.)
     The plate is then transferred to the 125° C char
block, and visualized by exposure to fuming sulfuric
acid (30 percent SO_3) for 20 min. Additional details
on layer equilibration, chromatogram development,
layer drying, and fuming sulfuric acid vapor char have
previously been published (5).

     Densitometry

Following visualization, the 25 chromatograms are densitometered as previously described (6). The basic
data handling and report generation procedures are
similar to those discussed in the chapter on source
rock analysis (7).

     RESULTS AND DISCUSSION

The plate of chromatograms selected to demonstrate
this method uses one standard to evaluate a wide range
of sample types. It should be noted that a different
standard could be prepared for each sample category
analyzed, and extremely accurate results obtained by
this method. However, as that point was illustrated
earlier (7), the utility of this basic TLC system
(5, 6) as a screening/fingerprinting tool is emphasized in this chapter.
     The first development moves all except the most polar
materials away from the sample application point. The second
development separates the aromatic and nonaromatic hydrocarbons
from the more polar compounds. The final development further
separates the aromatic and nonaromatic hydrocarbons, and also
helps to enhance resolution in the aromatic zone.
     As indicated in Figure 1, four distinct compound-
type zones are present in these chromatograms. The
nonaromatic hydrocarbons (NAH) are at the top of the
chromatograms. This zone contains the alkenes and

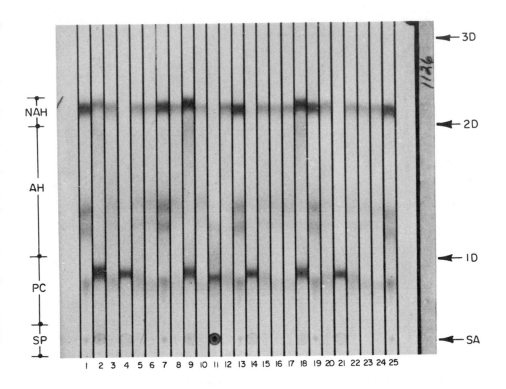

Figure 1. Plate No. 1126--the visualized chromatograms. The standard sample is on track numbers 1, 7,
13, 19, and 25; vacuum resids, tracks 2, 9, and·18;
flashed thermal tar, tracks 3, 12, and 20; pyrolysis
tar, tracks 4, 14, and 21; coker charge, tracks 5, 10,
15, and 22; heavy premium coker gas oil, tracks 6, 16,
and 23; hydrotreated gas oil, tracks 8, 17, and 24;
and pitch, track 11. (NAH, nonaromatic hydrocarbons;
AH, aromatic hydrocarbons; PC, polar compounds; and SP,
spot point material. SA is sample application point,
1D represents the extent of solvent migration in the
first development, 2D second development, and 3D third
development.)

alkanes in the samples. The next zone down on the
chromatograms contains the aromatic hydrocarbons (AH),
followed by the polar compound (PC) zone, which contains resins, most of the nitrogen, sulfur, and oxygen

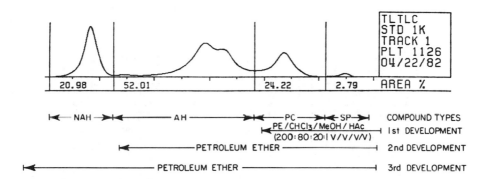

Figure 2. Computer plot of standard chromatogram on plate 1126, track 1. This figure summarizes the three developments and the resulting compound-type separations.

(NSO) species, and some pentane-insoluble asphaltenes. The fourth compound-type zone is designated the spot point material (SP), and contains the remainder of the pentane-insoluble asphaltene fraction, and any other very polar and/or highly insoluble components of the sample. The computer-generated plot of chromatogram number 1 from this plate graphically illustrates these four compound type zones (Figure 2).

    The computer-generated report is much the same as previously described (7) except that four compound type zones are determined rather than three. Table I contains some of the raw data for plate No. 1126. Column 1 is the track number, and columns 2 and 6 are the scaling points. Columns 3 through 5 are the so-called "scratch points" (the data points which are used to define the chromatogram compound-type zones, and which are selected based on slope change or valley location within a predetermined window). The sample scratch point locations are determined by extrapolation from the scratch points of adjacent standard samples (track numbers 1, 7, 13, 19, and 25). Columns 8 through 11 contain the net peak areas for the NAH, AH, PC, and SP compound types.

    Table II contains the normalized data. Column 1 is the sample or standard identity, column 2 is the sponsor's initials, column 3 is the plate number, and

TABLE I. PLATE 1126:   RAW AREA DATA

```
PLATE NO.  1126
 1    8 126 382 513 568 611   24.6702   46.4652   18.1897    1.1056
 2   24 125 381 512 573 611   16.3954   28.9862   32.0930    5.4692
 3   21 124 381 512 567 611   13.6258   31.2232   12.6376    0.1246
 4   41 123 380 511 571 611    1.4988   13.1414   20.2141    0.7591
 5    3 122 380.511 568 611   12.0270   31.0569   12.9659    0.2920
 6   33 121 379 510 567 611    9.2772   25.8803    8.3699   -0.3528
 7    9 120 379 510 569 611   24.8241   53.5089   19.8250    1.4096
 8   23 120 379 510 567 611   11.8950   25.5667    5.8866   -0.4071
 9   17 121 380 511 573 611   28.7150   35.7134   27.7445    3.1975
10    3 121 380 511 567 611    9.8514   26.0298    7.6257   -0.2689
11   35 122 381 512 576 611    1.2127   28.5758   34.0301   12.4586
12    7 123 381 513 567 611   15.7065   34.5810   13.4663    0.1598
13   15 124 382 514 569 611   25.7977   55.0393   20.0511    1.3077
14   35 123 382 513 570 611    2.8220   16.4429   21.0447    0.9723
15   20 123 382 513 567 611   12.4159   32.8806   13.5359    0.1923
16    3 123 382 512 567 611   10.7207   25.8604    7.3613   -0.1919
17   20 123 382 512 567 611   11.6239   24.9836    5.5615   -0.2879
18   10 123'382 511 572 611   27.4931   33.3871   25.6498    2.2532
19   11 123 382 511 569 611   23.4203   48.6366   19.0323    1.2811
20   29 124 383 511 567 611   16.6447   36.0105   13.6807    0.1230
21    3 125 385 511 572 611    2.0647   14.0448   20.2384    0.7090
22   25 126 386 511 567 611   12.3209   32.1868   14.0950    0.1302
23   27 127 388 512 567 611   10.7223   26.0396    8.7714    0.1198
24   22 128 390 512 567 611   11.8540   24.6352    5.5923   -0.2492
25    3 129 391 512 570 611   24.2912   48.1421   18.6767    1.0912
```

column 4 is the track number. Columns 5 through 8
contain the normalized peak areas for the NAH, AH, PC,
and SP compound types.   Column 9 contains the total
peak areas (TPA) values for each track.   Statistics
for the standard sample are shown at the top of this
table.

Table III is the report sheet used when 20 different samples are analyzed on one plate.   These data
were obtained in the same manner as those reported in
Table 4 of Ref. 7.   The only significant format differences are the inclusion of the sponsor's initials
(column 2), and the presence of four compound-type
results instead of three (columns 5 through 8).   As
will be noted on the first line under "Standard Statistics (n = 8)," the known standard sample composition
(in area percent) was NAH, 20.0 percent; AH, 53.0 percent; PC, 24.0 percent; and SP, 3.0 percent, with the
reasonable maximum TPA for the standard sample being
selected as 90.0 arbitrary units per track.   The geometric mean of the coefficients of variation for the
standard compound type results obtained by this method
is normally below 3 percent.

Figures 3 through 6 show the plots of the chromatograms for the standards and samples applied on the

TABLE II. PLATE 1126: NORMALIZED DATA

STANDARD STATISTICS

MEAN =			25.83	52.77	20.10	1.30	95.35
REL. SD.(%) =			3.747	1.983	1.836	7.792	5.442
STD 1K	1126	1	27.28	51.38	20.11	1.22	90.43
82737K LGB	1126	2	19.77	34.95	38.69	6.59	82.94
82780K DRP	1126	3	23.65	54.20	21.94	0.22	57.61
82781K DRP	1126	4	4.21	36.90	56.76	2.13	35.61
82782K DRP	1126	5	21.35	55.12	23.01	0.52	56.34
82783K DRP	1126	6	21.31	59.46	19.23	0.00	43.53
STD 1K	1126	7	24.93	53.74	19.91	1.42	99.57
82784K DRP	1126	8	27.44	58.98	13.58	0.00	43.35
82785K DRP	1126	9	30.11	37.45	29.09	3.35	95.37
82783K DRP	1126	10	22.64	59.83	17.53	0.00	43.51
81241L LGB	1126	11	1.59	37.46	44.61	16.33	76.28
82824K DRP	1126	12	24.57	54.11	21.07	0.25	63.91
STD 1K	1126	13	25.24	53.86	19.62	1.28	102.20
82825K DRP	1126	14	6.84	39.83	50.98	2.36	41.28
82826K DRP	1126	15	21.04	55.71	22.93	0.33	59.02
82827K DRP	1126	16	24.40	58.85	16.75	0.00	43.94
82828K DRP	1126	17	27.57	59.25	13.19	0.00	42.17
82829K DRP	1126	18	30.97	37.61	28.89	2.54	88.78
STD 1K	1126	19	25.35	52.65	20.60	1.39	92.37
82871K DRP	1126	20	25.05	54.18	20.59	0.19	66.46
82872K DRP	1126	21	5.57	37.90	54.61	1.91	37.06
82873K DRP	1126	22	20.98	54.80	24.00	0.22	58.73
82874K DRP	1126	23	23.49	57.04	19.21	0.26	45.65
82875K DRP	1126	24	28.17	58.54	13.29	0.00	42.08
STD 1K	1126	25	26.35	52.21	20.26	1.18	92.20

first 24 tracks of this plate. The sample types these
plots represent are listed in the legend of Figure 1.
Qualitative differences in the various sample compound-
type zones are readily apparent in the sample plots
(Figures 3 through 6). The small peaks marked with an
* in the figures (Figure 3, track 2; Figure 4, track 9;
and Figure 5, track 18) suggest the presence of certain naphthenes, olefins, and/or alkylbenzenes. The
trace for track 11 (Figure 4) dramatically demonstrates the result of using extrapolation between adjacent standards to divide the compound-type zones of
very dissimilar samples.

Fuming SO_3 char is an excellent visualization
technique, even though it is rather nonselective (8).
Although beyond the scope of the basic TLC system
under discussion (5, 6), these heavy refined oil sample chromatograms are certainly amenable to visualization while exposed to ultraviolet (UV) light prior to
visualization by char. A photograph of a plate visualized while exposed to UV light after the second development in petroleum ether is shown in Figure 7.

TABLE III. PLATE 1126: FINAL REPORT SHEET
FOR SINGLE ANALYSES

```
TLTLC PLATE NO. 1126                                    04/22/82
                              DATA CONVERSION FACTORS
GROUP 1 TRACKS   1- 7    0.7661 1.0083 1.1992 2.2742 0.9474
GROUP 2 TRACKS   7-13    0.7972 0.9851 1.2142 2.2261 0.8921
GROUP 3 TRACKS  13-19    0.7905 0.9952 1.1933 2.2501 0.9251
GROUP 4 TRACKS  19-25    0.7737 1.0108 1.1747 2.3343 0.9752

                              STANDARD STATISTICS (N=8)
ARTH. MEAN                20.001  52.999  24.000   3.000  90.000
STD. DEV.                  0.579   0.650   0.334   0.193   3.417
REL. SD. (%)               2.892   1.226   1.392   6.446   3.796
                         ***** GEOM. MEAN =   2.4% *****

  SAMPLE        PLATE  TK   NAH     AH      PC      SP      TPA
STD 1K           1126   1  20.981  52.012  24.216   2.791  85.672
82737K LGB       1126   2  13.548  31.525  41.512  13.415  78.579
82780K DRP       1126   3  18.198  54.887  26.421   0.494  54.579
82781K DRP       1126   4   2.844  32.826  60.053   4.277  33.739
82782K DRP       1126   5  16.238  55.189  27.403   1.170  53.377
82783K DRP       1126   6  16.436  60.351  23.213   0.0    41.237
STD 1K           1126   7  19.027  53.980  23.786   3.207  94.328


STD 1K           1126   7  19.847  52.865  24.141   3.147  88.828
82784K DRP       1126   8  22.677  60.231  17.093   0.0    38.672
82785K DRP       1126   9  23.151  35.581  34.069   7.198  85.083
82783K DRP       1126  10  18.368  59.976  21.656   0.0    38.814
81241L LGB       1126  11   0.985  28.676  42.090  28.250  68.049
82824K DRP       1126  12  19.782  53.823  25.833   0.562  57.019
STD 1K           1126  13  20.154  53.135  23.859   2.853  91.172


STD 1K           1126  13  19.987  53.681  23.449   2.884  94.545
82825K DRP       1126  14   4.861  35.655  54.717   4.767  38.191
82826K DRP       1126  15  16.601  55.347  27.320   0.732  54.606
82827K DRP       1126  16  19.712  59.858  20.430   0.0    40.653
82828K DRP       1126  17  22.584  61.106  16.310   0.0    39.012
82829K DRP       1126  18  23.979  36.659  33.769   5.594  82.136
STD 1K           1126  19  20.013  52.321  24.549   3.116  85.455


STD 1K           1126  19  19.562  53.073  24.137   3.228  90.082
82871K DRP       1126  20  19.620  55.457  24.485   0.437  64.813
82872K DRP       1126  21   3.875  34.438  57.672   4.015  36.139
82873K DRP       1126  22  16.176  55.210  28.098   0.516  57.278
82874K DRP       1126  23  18.353  58.232  22.796   0.619  44.522
82875K DRP       1126  24  22.566  61.270  16.164   0.0    41.039
STD 1K           1126  25  20.441  52.926  23.863   2.770  89.918
```

Although the black and white reproduction does not
give any indication of the various colors involved in
these chromatograms (green, orange, blue, purple, pink,
yellow), suffice it to say that a photograph, and in
particular a color photograph, can be a valuable

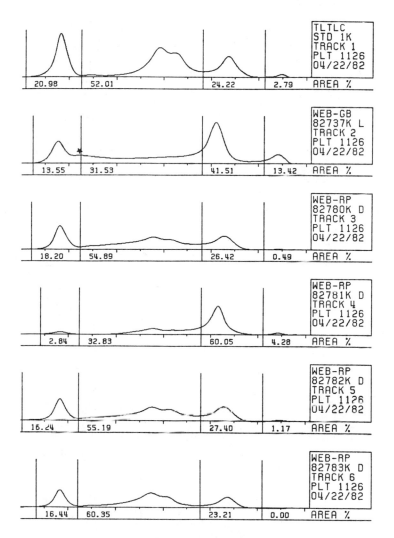

Figure 3. Computer plots of chromatograms on plate
1126, tracks 1-6. Sample identities are in the legend
of Figure 1.

supplement to the computer-processed data obtained following char visualization. Scanning these chromatograms before char while exposed to various wavelengths

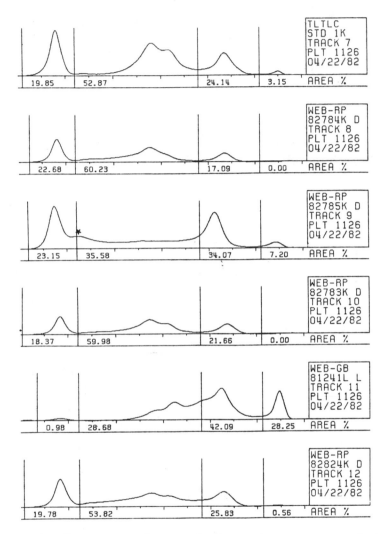

Figure 4. Computer plots of chromatograms on plate
1126, tracks 7-12. Sample identities are in the legend of Figure 1.

of ultraviolet light is another means of obtaining
data.   The fluorescence data obtained by using different filters and/or wavelengths of light can be

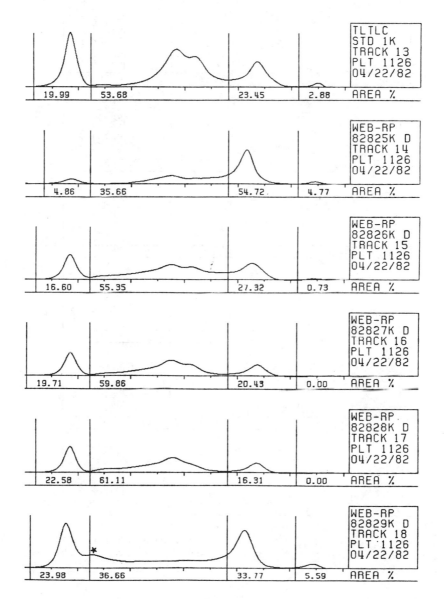

Figure 5. Computer plots of chromatograms on plate
1126, tracks 13-18. Sample identities are in the legend of Figure 1.

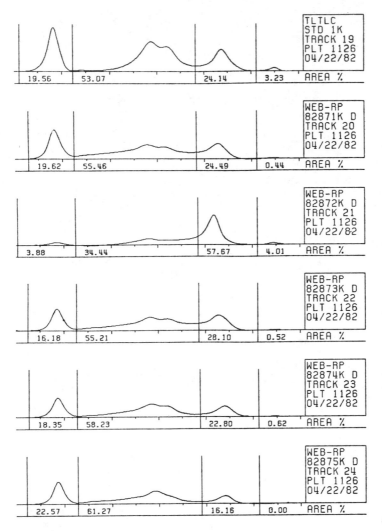

Figure 6. Computer plots of chromatograms on plate
1126, tracks 19-24. Sample identities are in the legend of Figure 1.

correlated with the transmittance data obtained after
char visualization, and thus provide additional information about the compounds present in the charred
spots.

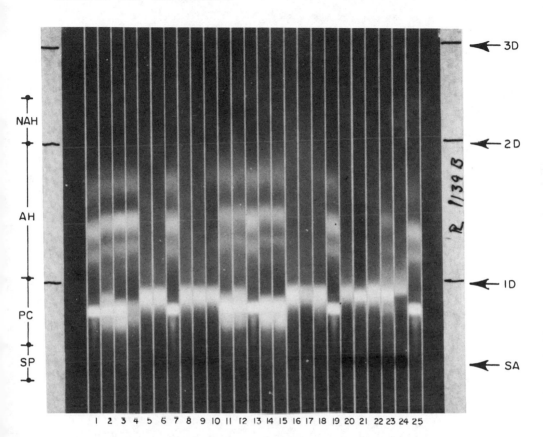

Figure 7. Photograph of plate 1139 chromatograms
(while induced to fluoresce by exposure to long wave
ultraviolet light). The standard sample is on track
numbers 1, 7, 13, 19, and 25; coker charge, tracks 2
and 12; heavy premium coker gas oil, tracks 3 and 14;
hydrotreated gas oil, tracks 4 and 15; vacuum resids,
tracks 5, 6, 8, 9, 10, 16, 17, and 18; thermal tar,
track 11; and ASCO bottoms, tracks 20 through 24.
This photograph was taken with a Nikon F3 camera while
illuminated by 366-nm UV light. Photograph conditions:
lens, Micro-NIKKOR 55 mm 1:2.8; settings, F4, 30 sec;
Kodak Wratten and Color Compensating filters, 2E, 81EF,
and CC30G; film, Kodak Pan-X (FX-135-20).

OTHER ANALYSES

The nature of the samples analyzed often makes compound-type determinations a very desirable type of
analysis. This basic TLC system (5, 6) has been
extended to include the quantitative determination of
sulfonates (9), diols (10), and sebum constituents (11)
as well as numerous other compound types. However,
this TLC system is equally suitable for quantitating
individual compounds in mixtures, such as additives in
grease, oils, or plastics.

CONCLUSIONS

This chapter shows how a basic TLC system (5, 6) may
be used by an experienced technician to obtain 75 to
125 analyses per day while determing the compound-type
distribution (nonaromatic hydrocarbons, aromatic hydrocarbons, polar compounds, and spot point materials) in
small stabilized samples of various coker feedstocks
and other refined petroleum fractions. The method is
presented and a set of samples taken through the
method, computer data processing, and report generation. The value of the results and of the computer
plots for screening/fingerprinting samples was also
noted.
       This series of four chapters has afforded us the
opportunity to describe in detail some of our techniques employed in layer, track, and chromatogram
preparation for quantitation (5) as well as a description of our automated TLC densitometry system and computer data handling procedures (6). A description of
a method used to characterize rock and soil extracts
for compound types provided the opportunity to see
these systems function to produce quantitative analytical results with particular emphasis on automated densitometry and computer data processing (7). This
final chapter demonstrates that the same system is
adaptable to other methods and sample types and can be
used as a semiquantitative screening tool. The sample
diversity in the petroleum industry seems to offer an
endless number of applications for TLC.

ACKNOWLEDGMENTS

For their contributions to the success of the TLC
effort at Conoco Inc., we wish to thank P. L. Smith,
R. L. Squibb, J. F. Stiverson, D. B. Summers, R. D.
Swain, R. M. Tillman, J. C. Touchstone, B. R. Warren,
G. D. Webb, S. M. Webb, N. O. Wolf, K. Yang, W. E.
Zabriskie, W. E. Zimmerman, and others.

REFERENCES

1. R. Amos, Chromatography in Petroleum Analysis,
   K. H. Altgelt and T. H. Gouw (Eds.), Dekker, New
   York, 1979, pp. 329-366.
2. H. J. Wernicke and E. Lassman, Compendium Deutsche
   Gesellschaft für Mineralölwissenschaft und Kohle-
   chemie 78-79(2), 1484-1498 (1978).
3. J. E. Ray, K. M. Oliver, and J. C. Wainwright,
   "The Application of the Iatroscan TLC Technique to
   the Analysis of Fossil Fuels," Petroanalysis '81,
   G. B. Crump (Ed.), Wiley, New York, in press.
4. M. L. Selucky, Anal. Chem. 55, 143 (1983).
5. J. B. Sudbury, T. T. Martin, and M. C. Allen, in
   Techniques and Applications of Thin Layer Chromatography, J. C. Touchstone and J. Sherma (Eds.),
   Wiley, New York, 1984, Chapter 11.
6. T. T. Martin, M. C. Allen, and J. B. Sudbury, in
   Techniques and Applications of Thin Layer Chromatography, J. C. Touchstone and J. Sherma (Eds.),
   Wiley, New York, 1984, Chapter 12.
7. T. T. Martin, M. C. Allen, and J. B. Sudbury, in
   Techniques and Applications of Thin Layer Chromatography, J. C. Touchstone and J. Sherma (Eds.),
   Wiley, New York, 1984, Chapter 13.
8. T. Jupille, J. Chromatogr. Sci. 17, 160 (1979).
9. M. C. Allen and T. T. Martin, JAOCS 48, 790 (1971).
10. T. T. Martin, Preprints, Div. of Petrol. Chem.,
    ACS 18, 562 (1973).
11. M. C. Allen, Tenside Detergents 11, 98 (1974).

CHAPTER 15

Thin Layer Chromatography— Densitometric Determination of Nitrite in Saliva

Joseph Sherma
Adelaide P. Lee
Mary S. Bowker
Albert J. Kind

INTRODUCTION

Because of the toxicological importance of nitrite ion
(1) and its possible relationship to the formation of
carcinogenic nitrosamines (2), a simple and selective
procedure for the quantification of nitrite in biological fluids is required. The usual methods for quantitative determination of nitrite in various samples
involve solution spectrophotometry after a color-form-
ing reaction (1-5) or ultraviolet spectrometry (6).
Since these methods tend to be more or less nonselective (7), a chromatographic method including separation of nitrite prior to determination is desirable.
Paper and thin layer chromatography (TLC) have been
widely used to qualitatively separate nitrite ions
from other anions (e.g., 8-15). Although the possibility of nitrite quantification by TLC with densitometry
has been inferred (16), successful application of this
technique has not been reported. This chapter
describes a system for separation, detection, and densitometric scanning of nitrite ions on preadsorbent
high-performance silica gel layers and demonstrates
applicability to the analysis of human saliva.

EXPERIMENTAL

Whatman LKDFHP precoated 60 Å high-performance plates
(10 x 10 cm) with a 2.1-cm diatomaceous earth sample
application area (the preadsorbent) and nine 8-mm
scored lanes were used. Plates were not prewashed but
were used as received.

A 50.0 ng/µl solution of nitrite was prepared by
dissolving 0.750 g of $NaNO_2$ in 100 ml of distilled
water and making a further 1.00:100 dilution of this
solution with water. Samples of 1.00-20.0 µl (50.0-
1000 ng) were applied by streaking across the preadsorbent area of the lanes using a 25-µl Drummond Diala-
matic microdispenser followed by drying with a low
intensity stream of warm air from a hair dryer (17).

To prepare the developing solvent, pyridine,
water, n-butanol, and concentrated ammonium hydroxide
were mixed in the volume ratio 20:40:40:5 in a separatory funnel. The mixture was shaken vigorously with
frequent venting for 5 min, and the layers were
allowed to separate. The bottom layer was discarded
and the upper (organic) layer was used as the mobile
phase.

Plates were developed immediately in a tightly
covered glass or metal N-chamber that was lined with
filter paper and equilibrated with a 3-mm pool of
mobile phase for 10 min before insertion of the spotted layer. Development was carried out within 1.5 cm
of the top of the plate. The chromatogram was removed
from the chamber and solvent was completely eliminated
using the hair dryer with the lowest heat/blower-
intensity settings (absence of solvent odor). All
operations involving pyridine should be carried out in
an efficient fume hood.

The following two reagent solutions were prepared:
(A) 0.50 g of sulfanilic acid in 100 ml of 30 percent
(v/v) glacial acetic acid (low heat is required to dissolve the sulfanilic acid completely), and (B) 0.10 g
of N-1-(naphthyl)ethylenediamine dihydrochloride in
100 ml of 30 percent (v/v) glacial acetic acid.
Nitrite was detected by spraying the layer with (A)
until the plate appears just to be soaked, and then
with (B) until red-purple bands appear. After drying
for 30 sec with a warm air stream, the plate was taped
layer-side down in the center of a clean 20 x 20 cm
glass plate.

The nitrite zone in each lane was scanned with a
Kontes Chromaflex fiber optics densitometer (K-495000)
in the single-beam, transmission mode using the 5-mm
scanning head and the visible wavelengths emitted by
the longwave UV source. The scanner was equipped with
a Kontes baseline corrector and Honeywell Electronik
194 recorder. Attenuation settings were between 50
and 200, the scan speed was 4 cm/min, and the recorder
speed was 1 cm/10 sec. Peak areas were measured as
height × width at half-height. Calibration curves
were plotted as area (mm^2) vs. ng of nitrite applied.
Areas were normalized for any attenuation changes.

Saliva was analyzed by direct application of samples to the preadsorbent. Standards were applied to
adjacent lanes in amounts that would bracket the
nitrite concentration expected in the sample. Saliva
was collected in a small tube or beaker by expectora-
tion without stimulation 5 min after rinsing the mouth
with water. The sample was refrigerated until use,
mixed well (e.g., Vortex mixing), and duplicate 20-30
µl aliquots were applied to the HPTLC plate. After
chromatography, detection, and zone scanning, a calibration curve was determined by computerized linear
regression of the areas of the standard peaks. Sample
concentrations were determined by computer fitting of
the average nitrite zone area to the standard curve.

RESULTS AND DISCUSSION

A large variety of brands and types of precoated
plates were tested, including fibrous and microcrystal-
line cellulose, acetylated cellulose, Ecteola cellulose, polyamide, aluminum oxide, silica gel, and chemically bonded C_{18} reversed-phase silica gel. Whatman
laned preadsorbent silica gel plates were chosen based
on R_f value, resolution, and tightness of zones; sensitivity of detection; and convenience of sample application and scanning.

It is important to dry standards and samples completely with a gentle stream of warm air prior to
development. Drying in an oven or with a stream of
hot air led to results that were less consistent, possibly due to heat lability of nitrite ions.

After testing numerous mobile phases suggested in
the literature and others designed in our laboratory,

pyridine-water n-butanol-ammonium hydroxide
(20:40:40:5 vv) was chosen. Development to within
1.5 cm of the top of the plate required about 60 min
and produced nitrite zones in the shape of bands
across the channels with an R_f value of 0.65. This is
within the ideal range for accurate and precise quantification as defined by Touchstone (18). Many of the
other mobile phases tested yielded more diffuse zones
and a secondary solvent front (after detection) close
to the position of the analyte. The mobile phase was
prepared fresh from the purest available reagents and
was used immediately.

The sulfanilic acid/(naphthyl)ethylenediamine
reagent (Marshall's reagent) was suggested by earlier
solution colorimetric analyses and was chosen after
evaluation of many other reagents, including other
formulations of these two chemicals, naphthylamine/
sulfanilic acid, $KI/AgNO_3$, Al(III)-morin complex, and
diphenylamine. The selected reagent produced an
intense red-purple zone that was shown to be stable
for up to 75 min if the chromatogram was covered with
a glass plate immediately after mobile phase evaporation. Stability was proved when a 500-ng zone was
scanned at 15-min intervals without any significant
variation in the resultant peak areas.

The minimum detection level for nitrite was 50 ng.
Since up to 30 μl of saliva could be directly spotted
on the preadsorbent without causing streaking, the
minimum detection on a concentration basis was 1.67
ng/μl. It was attempted to improve this level by use
of nonhigh performance 20 x 20 cm preadsorbent silica
gel layers. Because of their wider preadsorbent layer,
as much as 40 μl of saliva could be applied without
leading to streaking of the nitrite zone. However,
the less compact zones produced on the nonhigh performance layers made detection and scanning of the 50-ng
zone less reliable and, therefore, no significant
improvement in sensitivity occurred.

It is necessary to spray the sulfanilic acid
reagent until the layer appears just soaked, otherwise
fainter and less stable orange nitrite zones result.
The (naphthyl)ethylenediamine reagent must not be
oversprayed or the red-purple color will begin to diffuse, causing broadened zones. Nitrate ions do not
lead to the formation of the red-purple dye with the
reagent, so the detection method is selective for
nitrite in the presence of nitrate.

The HPTLC plate was taped to a 20 x 20 cm glass plate before scanning. The large plate served as the required "cover plate" for the Kontes scanner and also minimized exposure of the layer to air, which would lead to fading of nitrite zones and darkening of the background.

Calibration curves were linear from 50 to 1000 ng of nitrite, with correlation coefficients generally above 0.99. Typical slope and y-intercept values were 440 mm^2/ng and 1.1 x 10^4 mm^2, respectively. Although slopes and intercepts were quite constant from plate to plate, standards were always developed in parallel with samples on the same plate to correct for any variation.

Reproducibility was evaluated by applying seven replicate 10-μl (500 ng) portions of standard to a single plate. After development and detection, the relative standard deviation of the scan areas was 6.5 percent. As another indication of precision, the median degree of agreement between duplicate saliva aliquots was 11 percent. This level of precision has been shown (19) to be typical for analytical methods applied at concentration levels of 1-50 ppm, as is the case for nitrite in saliva.

Application of the method was demonstrated by analysis of duplicate aliquots of various saliva samples. In most cases, 20-μl samples were used and the lowest level of standard applied was 100 ng. The minimum level quantifiable was then 5 ng/μl, and levels below this amount were reported as "<5ng/μl." As indicated above, as low as 1.67 ng/μl can be determined by application of a 50 ng standard and 30 μl aliquots of saliva, and in a few analyses this was done. Table I shows data obtained for a single subject, an apparently healthy female college student, monitored over a period of 5 days upon waking and after lunch, and for morning samples of other subjects. These values are similar to those reported by Tannenbaum et al. (2). Morning samples for all subjects ranged from 2.4 to 10.2 ng/μl, although some of the samples indicated as "<5 ng/μl" probably contained amounts below 2.4 ng/μl. For subject 1, the range was 5 to 10.2 ng/μl. There was no clear trend for subject 1 comparing morning levels to those after lunch. A pooled sample containing about equal volumes of saliva from seven college students contained 6.9 ng/μl.

TABLE I. AMOUNT OF NITRITE IN SALIVA (ng/µl)

	Morning	After Lunch
Subject 1		
Day 1	10.2	3.9
2	7.0	6.2
3	9.5	1.7
4	<5	8.6
5	<5	9.8
Subject 2	8.3	
Subject 3	<5	
Subject 4	<5	
Subject 5	<5	
Subject 6	<5	
Subject 7	8.0	
Subject 8	5.2	
Subject 9	7.5	
Subject 10	2.4	
Pooled sample		
(7 subjects)	6.9	

To evaluate the accuracy of the method, a standard addition recovery experiment was performed. A saliva sample that assayed at 200 ng/30 µl (6.7 ng/µl) was fortified with nitrite to double its concentration. The sample was reanalyzed and after correction for dilution during the spiking step, a value of 390 ng/30 µl was obtained. This represented 95 percent recovery of the added nitrite.

SUMMARY AND CONCLUSIONS

A simple and reliable method is reported for determination of nitrite in saliva using thin layer chromatography and densitometry. No preliminary extraction of the sample is required since direct application to preadsorbent thin layer plates is possible. Sensitivity, precision, and accuracy appear to be adequate for use of the method in clinical analyses.

A major advantage of the method is the inherent selectivity for nitrite due to the chromatographic

separation involved. The results reported for saliva
samples should be considered as a maximum nitrite
level since identity of the analyte was not confirmed
by an independent analytical method.

Because the purpose of this research was to
develop the analytical method and not to collect data,
only a small number of actual samples were analyzed to
demonstrate its application. In addition to the
saliva analyses shown in Table I, fingertip blood was
collected from an adult male subject after breakfast
as previously described (20). Thirty microliters of
serum were applied and the method performed as
described for saliva. No nitrite was detectable in
this sample.

The amount of saliva applied to the plate should
be varied, up to 30 µl, according to the nitrite level
of the sample so that the amount of nitrite spotted
lies within the calibration curve. If it is desired
to apply larger amounts of saliva to increase the sensitivity of the method, deproteinization by addition
of two volumes of methanol followed by filtration or
low-speed centrifugation may be desirable. The sample
can then be directly applied or dried under a stream
of nitrogen and reconstituted with water.

The general method also should be applicable to
the determination of nitrite in other biological samples such as semen, sweat, or urine, either with
direct spotting or after appropriate sample preparation procedures.

ACKNOWLEDGMENT

The authors thank Dr. Irwin L. Shapiro for initially
suggesting this research and Stevin Zorn for some valuable preliminary work on the project.

REFERENCES

1. T. N. Wegner, J. Dairy Sci. 55, 642 (1976).
2. S. R. Tannenbaum, A. J. Sinskey, M. Weisman, and
   W. Bishop, J. National Cancer Inst. 53, 79 (1974).
3. J. Gabbay, Y. Almog, M. Davidson, and A. E. Donagi,
   Analyst 102, 371 (1977).
4. M. Nakamura and A. Murata, Analyst 102, 476 (1977).

5. N. R. Schneider and R. A. Yeary, Amer. J. Vet. Res. 34, 133 (1979).
6. O. C. Zafiriou and M. B. True, Anal. Chim. Acta 92, 223 (1977).
7. R. B. Lew, Analyst 102, 476 (1977).
8. V. D. Canic, M. N. Turcic, M. B. Bugarski-Vojino-vic, and N. U. Perisic, Z. Anal. Chem. 229, 93 (1967).
9. T. Okumura and Y. Nishikawa, Bunseki Kagaku 25, 423 (1976).
10. T. Okumura, Talanta 26, 171 (1979).
11. H. Thielemann, Z. Anal. Chem. 279, 365 (1976).
12. M. Subbaiyan and P. B. Janardhan, Proc. Indian Acad. Sci. 87A, 199 (1978).
13. R. Gallego, J. L. Bernal, and A. Martinez, Quim. Anal. 31, 69 (1977).
14. T. Okumura, K. Hiraki, and Y. Nishikawa, Bunseki Kagaku 26, 582 (1977).
15. J. Zuanon Netto, A. Longo, and L. W. Hanai, An. Farm. Quim. Sao Paulo 18, 103 (1978).
16. J. Franc and E. Kosikova, J. Chromatogr. 187, 462 (1980).
17. J. Sherma, Amer. Lab. 10(10), 105 (1978).
18. J. C. Touchstone, M. F. Schwartz, and S. S. Levin, J. Chromatogr. Sci. 15, 528 (1977).
19. W. Horwitz, "Analytical Measurements," in The Pesticide Chemist and Modern Technology, American Chemical Society, Washington, D. C., 1981, chapter 24, page 411.
20. J. C. Touchstone, G. J. Hansen, C. M. Zelop, and J. Sherma, "Quantitation of Cholesterol in Biological Fluids by TLC with Densitometry," in Advances in Thin Layer Chromatography: Clinical and Environmental Applications, J. C. Touchstone (Ed.), Wiley-Interscience, New York, 1982, chapter 16, page 219.

Practical Considerations in Quantitating Gangliosides by High-Performance Thin Layer Chromatography

Brian R. Mullin
Charles M. B. Poore
Bonnie H. Rupp

INTRODUCTION

With the development of high-performance thin layer chromatography (HPTLC) plates, densitometric quantitation of gangliosides on silica gel layers has become more sensitive and reproducible (1-3). This is a result of better resolution on the more efficient and uniform thin layer (4) as well as advanced instrumentation and improved chromatographic techniques. A methodology is discussed for the densitometric quantitation of gangliosides that is 1000 times more sensitive than spectrophotometric methods (5-7). Technical aspects of sample application, development, and scanning densitometry that enhance the sensitivity and reproducibility of ganglioside quantitation will undoubtedly be beneficial in the quantitative analysis of other analytes.

Gangliosides are a diversified family of glycosphingolipids characterized by the presence of sialic acid. They are present in high concentrations in the central nervous system of higher vertebrates (8). The sialic acid moiety specifically stains blue with resorcinol reagent (5) and can be quantitated by determining the absorbance at 580 nm. The five major

TABLE I. CHEMICAL STRUCTURES OF GANGLIOSIDES
UNDER STUDY

G_{M4}:   ceramide-gal
                        |
                      NANA

G_{M1}:   ceramide-glu-gal-galNAc-gal
                                      |
                                    NANA

G_{D1a}:   ceramide-glu-gal-galNAc-gal
                                | |
                              NANA   NANA

G_{D1b}:   ceramide-glu-gal-galNAc-gal
                                      |
                                    NANA
                                      |
                                    NANA

G_{T1b}:   ceramide-glu-gal-galNAc-gal
                                | |
                              NANA   NANA
                                |
                              NANA

Abbreviations: gal = galactose; glu = glucose;
galNAc = N-acetylgalactosamine; NANA = N-acetylneura-
minic acid or sialic acid. Ganglioside nomenclature
is that of Svennerholm (10).

gangliosides of human central nervous system have been
selected for study. Their chemical structures are
shown in Table I.

MATERIALS AND METHODS

Materials used have been previously described in
detail (1, 2).
        HPTLC plates were prewashed in chloroform-metha-
nol (2:1) and activated at 140° C for 10 min. Pure

gangliosides and ganglioside mixtures in chloroform-methanol (2:1 v/v) were applied to the plates in 3-mm bands in a volume of 1-5 µl with a microsyringe (Hamilton, 7000 series). Samples to be quantitated were diluted so that 1 to 5 µl aliquots fall within the range of standard curves (10 to 100 pmol ganglioside sialic acid). Unknown samples were applied between duplicate applications of gangliosides used for the standard curve on the same plate. A standard ganglioside mixture contained equal sialic acid amounts of G_{M4}, G_{M1}, G_{D1a}, G_{D1b}, and G_{T1b} in chloroform-methanol (2:1).

Four different mobile phases were initially tested for their ability to separate complex ganglioside mixtures:

Mobile Phase	Composition
A	chloroform-methanol-water-1% $CaCl_2$ (55:45:8:2)
B	chloroform-methanol-water-1% $CaCl_2$ (60:35:7:1)
C	chloroform-methanol-water (70:30:4)
D	chloroform-methanol-4,5N NH_4OH (60:35:8)

Mobile phases A and B were found to be the most effective in separating individual ganglioside species without causing excessive diffusion of the fast-moving ganglioside bands (G_{M4} and G_{M1}). Ganglioside mixtures were studied using mobile phases A and B developed to heights ranging from 80 to 155 mm.

Ganglioside bands were scanned with a Shimadzu dual-wavelength TLC scanning densitometer, Model CS-910. The transmission mode was selected because it produced twice the densitometer detector response compared with the reflection mode. The dual-wavelength capability of the scanner allowed simultaneous scanning of the same area at two different wavelengths. The sample wavelength was 580 nm, the absorbance maximum of the sialic acid-resorcinol chromophore, and the reference wavelength was 720 nm, at which ganglioside bands absorbed the least energy. The slit length was adjusted to be 10 percent greater than the widest band

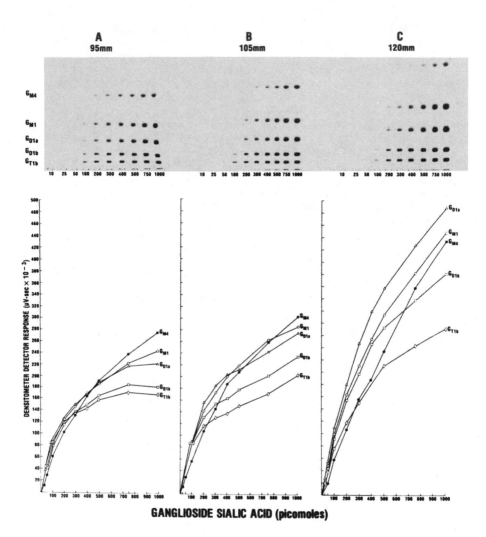

Figure 1. Standard curves for a ganglioside mixture
developed in mobile phase A, on 10 x 20 cm silica gel
60 HPTLC plates (E. Merck).   Each lane contains the
indicated pmol amounts of sialic acid for each ganglioside species.   Detector response is plotted as a function of amount of ganglioside sialic acid.   Detector
response is the peak area due to absorbance at 580 nm
minus absorbance at 720 nm.

after development, and the slit width was 0.2 mm.
Each lane was scanned at a rate of 6 mm min. The peak
area of each band was calculated after subtracting the
background absorbance at 720 nm (Shimadzu data processor, Model C-R1B). Neither the zigzag function nor
the linearizer program was used.

DISCUSSION

Necessity of Standard Curves in Densitometric Quantitation

The major problem in quantitating individual species
within a mixture by densitometry on HPTLC plates is
that equal amounts of the various species within a mixture do not necessarily cause the same densitometer
detector response after separation on the thin layer.
When a mixture of gangliosides is separated by HPTLC
and scanned, each ganglioside species has a different
standard curve (Figures 1 and 2). Since these differences are not observed when individual gangliosides
are applied in uniform bands and stained without development, it appears that differences in ganglioside
standard curves are a consequence of differing band
areas or geometry. Differences in detector responses
may also result from inconsistencies in the gel layer
and in spraying. Therefore, when quantitating individual ganglioside species within a mixture, it is manda-
tory that ganglioside standards for use in constructing standard curves be developed on the same plate.
    The extent of linearity of standard curves is
affected by both the composition of the mobile phase
and the height of the mobile phase front (Figures 1
and 2). Effective chromatographic conditions for quantitating ganglioside species within a mixture are
mobile phase B developed to a height of 120 mm (Figure
2).

Errors Involved in Densitometric Quantitation

Various steps can be taken to minimize errors in quantitating gangliosides by scanning densitometry on HP-
TLC plates. Samples applied with a positive-displace-
ment microsyringe (Hamilton, 1-5 μl) show <5 percent
variability, while those applied with a microcapillary

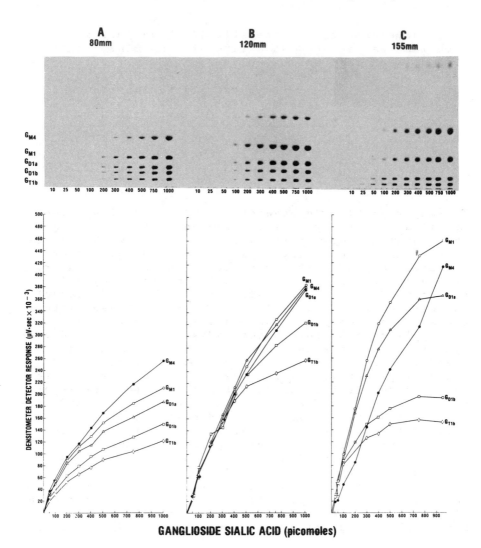

Figure 2. HPTLC plates are shown with corresponding
densitometric standard curves for a standard ganglioside mixture developed in mobile phase B. Each lane
contains the indicated pmol amounts of sialic acid for
each ganglioside species. The height above the origin
of mobile phase development is indicated for each
chromatogram.

TABLE II. DENSITOMETRIC QUANTITATION
OF SPOTS vs. BANDS

| Ganglioside | Percent Error | |
	Spots	Bands
G_{M4}	16	5
G_{M1}	3	5
G_{D1a}	6	3
G_{D1b}	6	4
G_{T1b}	7	4

Sialic acid (5, 10, 25, and 50 pmol) amounts of each
ganglioside were applied in quadruplicate 1 mm spots
on 10 x 10 cm HPTLC plates. Sialic acid (10, 20, 30,
40, and 50 pmol) amounts of each ganglioside were
applied in triplicate 3 mm bands on 10 x 10 cm HPTLC
plates. Each plate was developed in mobile phase B to
a height of 80 mm above the origin and quantitated by
scanning densitometry.

pipet (Wheaton, 1-5 µl) show 10 percent variability.
Automatic sample applicators have previously been
shown to have <5 percent variability (9).
    Ganglioside samples in 1-5 µl can be conveniently
applied in 3-mm bands. As little as 1 µl can be
evenly distributed over this distance in overlapping
spots which coalesce into a 1 x 3 mm band. Gangliosides applied in bands show a variability of <5 percent, while ganglioside spots vary as much as 16 percent (Table II). This confirms the observation of
Touchstone and Levin (9) that bands are superior to
spots. Samples applied in aqueous solvents produce
larger, more irregular bands than do samples in anhy-
drous solvents. Chloroform-methanol (2:1 v/v) was
selected as the sample solvent because gangliosides
are highly soluble in this mixture, and solutions can
be applied to gels in small drops that yield narrow
sample bands. Furthermore, this solvent is easily

TABLE III. ERRORS ASSOCIATED WITH DENSITOMETRIC
QUANTITATION OF A SINGLE GANGLIOSIDE SPECIES

Step	Percent Error
Alignment and scanning	2.0
Sample application and staining	3.6
Plate-to-plate variability	4.2

Bands (10 and 50 pmol) of G_{M1} (>99 percent pure) were
applied 11 times each with a microsyringe on 10 x 10
cm HPTLC plates and developed in mobile phase B to a
height of 80 mm. Sample volumes were $\leq$5 µl. Each
band was scanned five times, and the mean densitometer
detector response and standard deviation were determined. Plate-to-plate (interassay) variability was
determined by quantitating a G_{M1} sample of unknown concentration on three different 10 x 10 cm HPTLC plates
(2).

evaporated from the gel layer with warm air and does
not cause streaking as some aqueous solvents.
     Errors associated with the various steps in the
densitometric quantitation of gangliosides are shown
in Table III. When a single stained ganglioside band
is aligned with the light slit and repeatedly scanned,
the variability is 2.0 percent. This value represents
the minimum error involved in this procedure and is
attributable to the instrumentation system alone.
Variability in sample application and staining is
approximately 4 percent using a microsyringe to apply
a 1-5 µl sample. Quantitation of a single pure ganglioside species using standards on the same plate has
a variability of <5 percent from one plate to another.
When gangliosides within a mixture are quantitated,
the plate-to-plate error increases slightly to about
6 percent (Table IV). This increased experimental
error apparently results from difficulty in slit alignment and fluctuations in the output of the source over
longer scanning distances. A slit length 10 percent
longer than the widest band was found to facilitate
alignment when quantitating ganglioside mixtures and
minimizes plate-to-plate variability.

TABLE IV. ERRORS ASSOCIATED WITH DENSITOMETRIC
QUANTITATION OF GANGLIOSIDES WITHIN A MIXTURE

Step	Percent Error
Sample application and staining	4.8
Plate-to-plate variability	5.7

Unknown mixtures containing three or more gangliosides
in the range of 25 to 100 pmol were applied in triplicate 3-mm bands on four different 10 x 20 cm HPTLC
plates. All plates contained 25-, 50-, 75-, and 100-
pmol amounts of a standard ganglioside mixture for densitometric standard curves. Plates were developed in
mobile phase B to a height of 120 mm above the origins
(1). Plate-to-plate (interassay) variability was
determined by quantitating each unknown ganglioside
species within the mixture on the four different
plates.

Methods to Increase Sensitivity

Sensitivity is defined as the minimum amount of ganglioside sialic acid that can be reproducibly detected
above background noise. Prewashing with chloroform-
methanol (2:1 v/v) and thorough removal of mobile
phase prior to resorcinol staining reduce background
interference. Use of the transmission mode gives
double the densitometer detector response compared to
the reflection mode. Simultaneous dual-wavelength
scanning, by compensating for background interference,
enhances sensitivity, as does slower scanning speeds.
Although narrowing the slit width will increase the
resolution of closely spaced bands, slits of 0.1 mm or
less will significantly decrease sensitivity by
increasing background noise. Using these conditions,
as little as 1 pmol of ganglioside G_{M1} can be reproducibly detected with a signal/noise ratio of 5:1
(Figure 3).
    The experimental conditions discussed will allow
for sensitive and reproducible quantitation of gangliosides, either pure or in mixtures, on HPTLC plates.

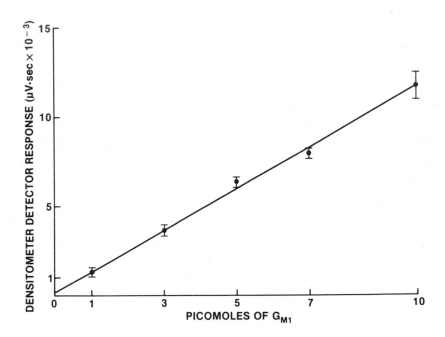

Figure 3. Limits of ganglioside detectability by scanning densitometry of HPTLC plates. G_{M1} was applied to a 10 x 10 cm HPTLC plate in 3-mm bands in the amounts indicated. The plate was developed in mobile phase B to a height of 80 mm above the origin. Densitometer detector response values are the mean +SD of triplicate samples (correlation coefficient = 0.987).

ACKNOWLEDGMENTS

This research was supported by a grant from the National Multiple Sclerosis Society.

REFERENCES

1. B. R. Mullin, C. M. B. Poore, and B. H. Rupp,
   J. Chromatogr., in press.
2. B. R. Mullin, C. M. B. Poore, and B. H. Rupp,
   J. Chromatogr., submitted June 1983.
3. S. Ando, N. C. Chang, and R. K. Yu, Anal. Biochem.
   89, 437 (1978).
4. J. Ripphahn and H. Halpaap, J. Chromatogr. 112, 81
   (1975).
5. L. Svennerholm, Biochim. Biophys. Acta 24, 604
   (1954).
6. L. Warren, J. Biol. Chem. 234, 1971 (1959).
7. D. Aminoff, Biochem. J. 81, 384 (1961).
8. F. B. Cochran, R. K. Yu, and R. W. Ledeen, J.
   Neurochem. 39, 773 (1982).
9. J. C. Touchstone and S. S. Levin, J. Liquid Chromatogr. 3, 1853 (1980).
10. L. Svennerholm, J. Neurochem. 10, 613 (1963).

CHAPTER 17

Analysis of
L-Alanine *N*-Carboxyanhydride
by Quantitative
Thin Layer Chromatography

Russell J. Lander
Laszlo R. Treiber

INTRODUCTION

As part of a process development effort, the synthesis
of L-alanine N-carboxyanhydride (NCA) from L-alanine
was investigated. The procedure was first developed
by Fuchs (1), Farthing (2), and Levy (3), wherein
milled amino acid is added to an excess of phosgene in
an inert solvent (dioxane, ethyl acetate) and agitated
at 30-50° C for several hours. The product is
stripped of excess phosgene by distillation and isolated via crystallization. The formation of L-alanine
NCA is illustrated in Figure 1.

To gain understanding of the reaction kinetics
and maximize the yield, a method for analysis of
L-alanine NCA was desired. The standard technique,
involving measurement of carbon dioxide generation on
acidification, was too cumbersome to accommodate the
number of samples generated by these experiments. The
thin layer chromatographic method was hindered by the
instability of alanine NCA (NCA's are moisture sensitive, hydrolyzing to the respective amino acids).
Reaction of alanine NCA with an excess of n-butylamine
afforded a stable derivative with strong absorbance at
210 nm (Figure 1).

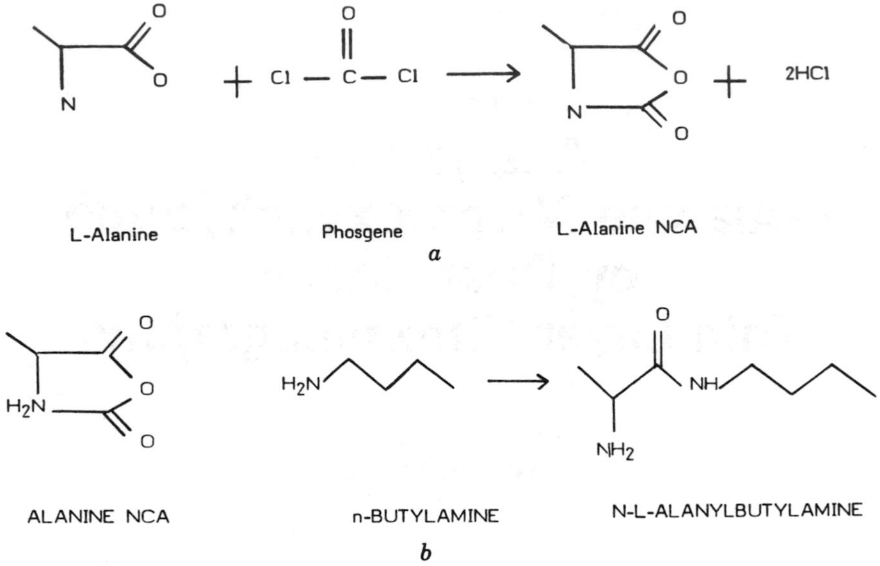

L-Alanine   Phosgene   L-Alanine NCA

a

ALANINE NCA   n-BUTYLAMINE   N-L-ALANYLBUTYLAMINE

b

Figure 1. Chemistry of alanine NCA. (a) Formation of alanine NCA. (b) Reaction of alanine NCA with n-butylamine.

        To prevent competition of phosgene, samples were evaporated and redissolved in tetrahydrofuran (THF). This procedure afforded a clean and highly reproducible chromatographic peak. The derivative is also soluble in water, and as such, could be stored as a stable standard solution for months.

        EXPERIMENTAL

L-alanine NCA synthesis was conducted in a THF solution containing about a sixfold excess of phosgene. Solid alanine was added as a THF slurry. Reaction completion was signaled by the consumption of the solid (solution clarity). Time samples for the kinetic study were filtered through a fully enclosed glass frit attached to the bottom outlet (to remove unreacted alanine on the frit).
        One ml of the clear filtrate, containing about 60 g L-alanine NCA, was evaporated to remove unreacted

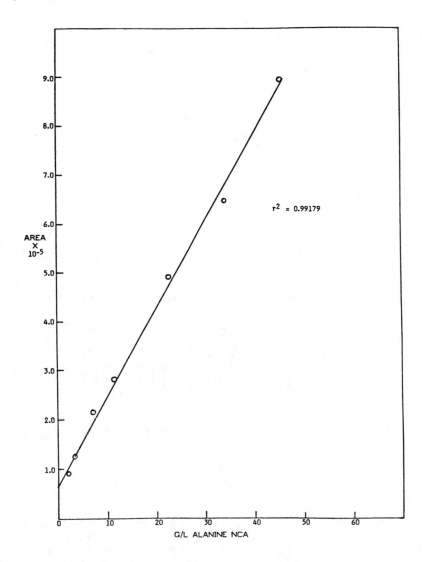

Figure 2. QTLC calibration: the best fitting straight
line calculated by the least squares method.

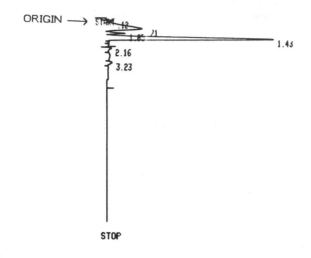

RUN #   37   DEC/02/82 10:20:08

AREA%
  RT   AREA TYPE   AR/HT   AREA%
  0.12   24751 D BV   0.087   1.245
  0.71   508810   VV   0.358   25.587
  1.05   81950   VV   0.112   4.121
  1.43   1312700   VV   0.192   66.011
  2.16   19071   BP   0.143   0.959
  3.23   41304   VV   0.223   2.077

Figure 3. QTLC scan of N-L-alanylbutylamine.

phosgene.   The resulting oil was dissolved in one ml
of THF.   One ml of 20 percent (v/v) n-butylamine in
THF was added rapidly, and the sample was manually
shaken for 1 min.

    Ten µl of sample (derivatized) was applied as a
spot to a silica gel plate (Kieselgel 60F-254 precoated TLC plates, E. Merck).   After drying for a half-
hour at room temperature, the plate was developed with
ethyl acetate-ethanol-28 percent ammonia (80:20:4 v/v).
The plate was dried at room temperature and scanned
with a Shimadzu CS-920 TLC scanner at 210 nm.

TABLE I. CORRELATION COEFFICIENT (r^2)
vs. PLATE POSITION AND INTEGRATOR TYPE

NORMAL POSITION		REVERSE POSITION	
H PACKARD	SHIMADZU	H PACKARD	SHIMADZU
0.99346	0.98801	0.99179	0.99268

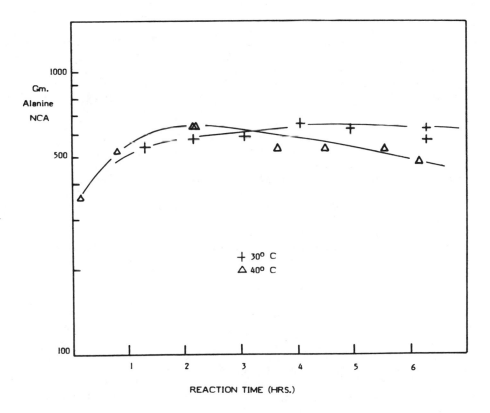

Figure 4. Kinetics of alanine NCA formation of 30° and 40° C.

RESULTS

Eight solutions of alanine NCA were prepared from a
crude reaction mass concentrate of 95 percent purity
(not crystallized). The solutions spanned a twenty-
fold range of dilutions, the maximum concentration
corresponding to the typical final reaction solution
condition. These were coupled with n-butylamine, separated by TLC, and scanned. The results were plotted
(Figure 2) and found satisfactory (r^2 = 0.99179, where
r = correlation coefficient). Note that the correlation embraces all experimental sources of error in the
derivatization, spotting, and scanning sequence.
     Solutions at 45 g/l alanine NCA were spotted on a
single plate, developed, and scanned. This procedure
was repeated with a solution close to the minimum
detection limit (5.9 g/l). Standard deviation at the
maximum concentration was $\pm$ 2.9 percent and at the low
concentration, $\pm$ 5.5 percent. These limits satisfied
the requirement of the kinetic study.
     Figure 3 shows a full scan of a single developed
chromatogram. Impurities close to the origin are
probably reaction by-products of butylamine and phosgene. Unreacted alanine NCA would hydrolyze to alanine, which has a weak absorbance at 210 nm. The
small peak close to alanine may be the result of
further reaction to N-L-alanylbutylamine with (unreacted) alanine NCA.
     For comparison, the plate containing the calibra-
tion samples (Figure 3) was scanned in a reverse mode,
starting from the clean, solvent front side. Table I
shows the correlation coefficient obtained for each
scanning direction using two different integration systems as well. Using the Shimadzu integrator,
correlation is improved by reversing plate direction.
The Hewlett-Packard (H.P.) integrator performs well
when scanning from either direction. The latter features an interpolated baseline correcting for baseline
drift, whereas the Shimadzu employs only a horizontal
baseline referenced at each peak. The difference in
the r^2 values, though small, is statistically significant.
     The derivatization/QTLC technique proved effec-
tive in the study of reaction kinetics for the formation of alanine NCA from L-alanine and phosgene (Figure 1). Figure 4 shows the kinetics of alanine NCA

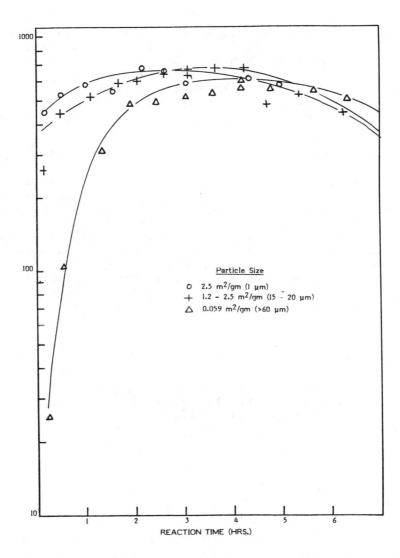

Figure 5. Kinetics of alanine NCA formation: three
particle sizes.

formation at 40 and 30º C. A consecutive degradation
reaction at 40º C is effectively eliminated by lowering the reaction temperature to 30º C. Figure 5 shows
rate data (at 40º C) for several alanine particle
sizes. Faster reaction with small particle size
brings about a yield increase by reducing losses to
consecutive degradation.

CONCLUSIONS

Quantitative TLC is demonstrated to be effective in
the analysis of L-alanine NCA by way of derivatization.
This method should find broader application in the synthesis of peptides where the use of NCA derivatives of
amino acids is a popular method of obtaining optically
pure compounds. Accuracy of the technique should be
sufficient even for these demanding multiple step
sequences. Moreover, the assay has been successfully
used to characterize the kinetics of formation of alanine NCA, affording, thereby, a relatively pure product.

REFERENCES

1. F. Fuchs, Ber. 55, 2943 (1922).
2. A. Farthing, J. Chem. Soc. 3213 (1950).
3. A. Levy, Nature 165, 152 (1950).

CHAPTER 18

Quantitative Aspects of Thin Layer Chromatography— FID of Polymers

Victoria L. Dimonie
Mohamed S. El-Aasser
John W. Vanderhoff

INTRODUCTION

A growing interest in "tailor-made" polymers as new polymeric materials has been recognized in recent years. The structural and chemical heterogeneity of such materials strongly influences their properties, which cannot be interpreted only by the average values of the polymer properties as obtained using usual analytical methods. Even in the case of statistical copolymers, the chemical composition as well as the monomeric arrangement (sequence length) in the chain and molecular weight distribution are important factors in characterizing the properties of the product. For more complicated systems such as graft-copolymers, not only the chemical composition but also the degree of grafting (the ratio of attendant homo- or copolymer and the ungrafted mother polymer) need to be known. Therefore, it is desirable to have a rapid and automatic method for the determination of chemical composition and structural heterogeneity of polymeric materials. All methods require separation of the polymer into different fractions depending on a particular polymer property, followed by analysis of the various fractions.

Several separation methods such as fractional precipitation and dissolution, density-gradient ultracentrifugation have been used to separate polymers. However, thin layer chromatography (TLC) was found to be an adequate and relatively simple technique to separate polymers.

Many reports appeared on the application of TLC for the characterization of polymers with respect to differences in chemical composition (1-5), extent of blockiness (6), monomer arrangement (7-10), steric tacticity (11-14), and molecular weight distribution (15-19). Important progress has been made in determining the purity of graft copolymers, and in isolating "truly" grafted copolymer species from a given reaction product. TLC has been effectively applied to test whether or not a graft product contains the attendant homopolymer and ungrafted mother polymer (20), or to determine the true graft efficiency (21, 22) that can be hardly found by usual isolation techniques.

The separation of the sample into several components and the qualitative identification of discrete spots is a relatively easy matter. More complex is the quantitation of chromatograms even when good separation among the spots is achieved. The quantitative analysis can be done either by photofilm recording of the spot darkness on visualized chromatograms (1, 5) or using a scanning spectrodensitometer (7-11, 18-21, 23). TLC scanning is possible only for the samples that have specific absorption in the ultraviolet and/ or visible region.

The Iatroscan TLC-FID analyzer offers a unique method of direct quantitation by combining the TLC technique with an automated quantitative detection system based on the classical GC/Flame ionization principle. In spite of the substantial number of papers dealing with the TLC-FID separation of different organic compounds with low or relatively low molecular weight, there are only few papers discussing the application of TLC-FID technique to polymers (24, 25).

The goal of this paper is to present some specific aspects of polymer separation on silica chromarods and examples of quantitative determination of polymer composition by TLC-FID.

EXPERIMENTAL

Polymer Samples

A series of polystyrene (PS) samples with very narrow
molecular weight distribution (standard samples for
GPC) were used. The styrene-acrylonitrile copolymers
(St-AN) with different compositions were prepared by
thermal polymerization with conversion of about 10
percent. The AN content of copolymers was determined
by nitrogen analysis. The styrene-methylmethacrylate-
ethylbutylacrylate (SMB) copolymer was supplied
through the courtesy of Iatron Laboratories Inc.

     For PS and St-AN copolymer samples the stock solu-
tions were prepared by dissolving the polymer in THF
at concentrations of 2 mg/ml. The SMB copolymer was
dissolved in chloroform at a concentration of 10 mg/ml.

TLC-FID Separation Technique

The Iatroscan TH-10 TLC/FID analyzer (Iatron Laboratories, Inc., Tokyo, Japan) was used for quantitative
chromatographic separation of the polymer. The separations are done on the TLC rods (Chromarods) in the
normal manner by solvent development. The chromarods
are quartz rods coated with a special sintered silica
gel of 5-μm particle size. Two types of chromarods
with different thickness of the silica layer were used:
S_{II} type rods with a 75-μm silica layer and S_{200} rods
with 200-μm silica gel.

     The following procedure was applied: The chroma-
rods were cleaned up and activated by blank scanning
the rods through the FID and observing the stability
of the recorder baseline. If it is necessary a second
FID scan is repeated until a stable baseline is
achieved, signifying zero-rod contamination. With the
aid of a microsyringe a spot of stock solution containing ca. 4 μg of polymer is formed on each rod, 3 mc
from the end. After spotting, the chromarods retained
in the rod holder are dried in the oven for 5 min.
The rod holder is transferred into the development
chamber, which is saturated with solvent vapors.

     The development is performed so that the solvent
front is located at the 10-cm position on the chroma-
rod. The rod holder is removed and placed in the oven
at 100° C for 5 min. If multidevelopments are

required, the rods are dried in the oven before the
next development. In the case of SMB copolymer, the
first development was performed with diethyl ether
for a distance of 10 cm and the second one with acetone for 5 cm.

    After drying, the rod holder is rapidly trans-
ferred from the oven to the scanning area of the TH-10
to avoid atmospheric contamination of the rod coating.
Output from the FID is recorded on a chart recorder.

RESULTS AND DISCUSSION

The main objective of this work was to characterize
the chemical composition and degree of heterogeneity
for St-AN copolymers prepared by emulsion polymerization processes.

    Based on the data published by Glockner and
coworkers (6), a good TLC separation can be obtained
on silica plates for St-AN copolymers with different
chemical compositions. Well-defined spots without
tailing were obtained for the St-AN copolymers with a
very narrow chemical composition and high molecular
weight using an elution gradient technique with tolu-
ene-acetone mixture. The R_f values increase with
increases in the St content of the copolymers. The
TLC-FID technique was used for quantitation of the
chemical heterogeneity of the St-AN copolymers. The
same development technique used on silica plates was
applied to the separation on chromarods type S_{II}.
Figure 1 shows that the resulting peaks after development and scanning are very broad; the broader the
peaks are, the higher the migration distance (R_f) of
the sample on the rod. Very poor reproducibility of
the peaks was obtained when the same polymer sample
was analyzed on different rods. Better reproducibility of the measurements was obtained for the copolymer
samples with a richer AN content, which did not
migrate so far from the starting point (Figure 2).

    Even in case of PS standard samples for GPC with
very narrow molecular weight distribution and using
single solvents as developing agents, without gradient
concentration development the recorded peaks for the
FID response are very broad (Figure 3), regardless of
the type of solvent used. These results are in contrast to the reported results of well-defined peaks

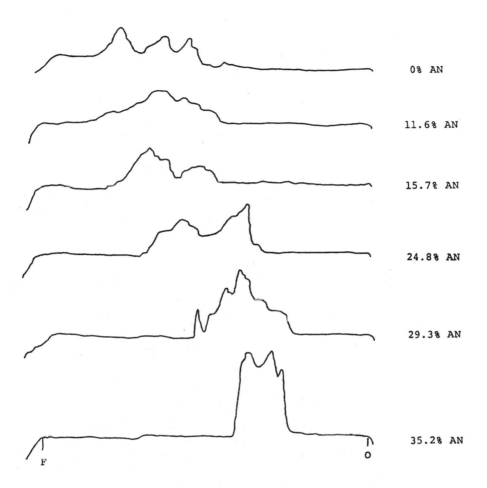

Figure 1. Chromatograms of St-AN copolymers eluted on chromarods S_{II} using the solvent gradient technique with acetone-toluene.

obtained for a large number of organic substances with low molecular weight. Consequently, it was concluded that the only difference that can affect the separation results is the higher molecular weight of the polymer samples. In order to understand the reasons that can cause such a broadening of the peaks in the

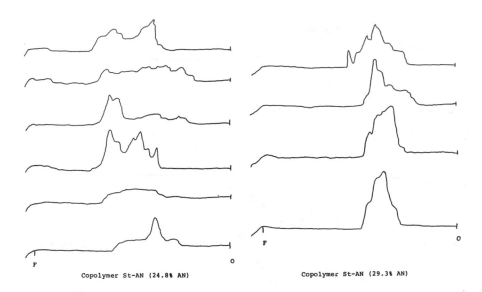

Figure 2. The reproducibility of the FID response on
S_{II} chromarods developed at the same time.

case of polymers, two PS samples with different molecular weights were developed using benzene on two types
of chromarods, S_{II} and S_{200}. One of these PS samples
has a low molecular weight (2350), close to that of
lipids which were successfully separated on chromarods.
The other is a PS sample with much higher molecular
weight (ca. 110,000), but within the limits of average
values for a large number of polymers.

Figure 4 shows the FID responses after development with benzene of both polystyrene samples on the
two rods with different thickness of the silica coating. For the S_{II} type rods, which have a thinner silica layer, a sharp peak was recorded only for the low
molecular weight PS sample. In the case of S_{200} rods,
which have a three-times thicker silica coating than
that of S_{II} rods, well-defined peaks were obtained for
both samples.

The data presented in Table I on the FID response
for the PS sample on both types of rods show that the
areas under the peaks are always smaller for the

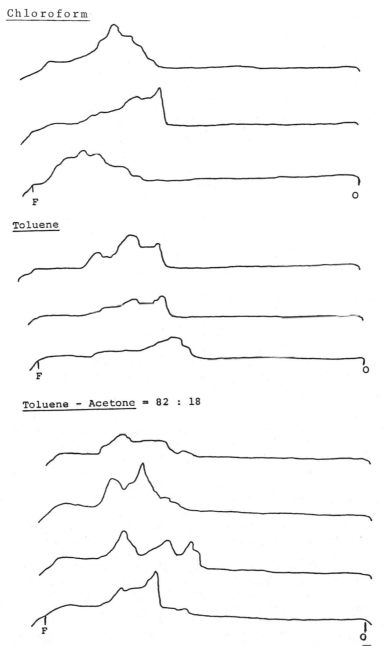

Figure 3. The chromatograms for a PS sample ($\overline{M}_n$ = 200,000) on chromarods S_{II} using different solvents for the development.

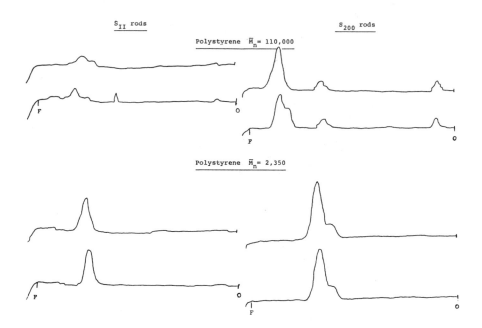

Figure 4. The chromatograms for PS samples of two different molecular weights developed with benzene on S_{II} and S_{200} chromarods

eluted samples than that obtained for the same amount of spotted sample but without development. The difference is larger, the higher the molecular weight of the polymer. For the same PS sample lower values for the FID response are obtained in the case of S_{II} rods compared to S_{200} rods. When S_{200} rods are used, even at higher amount of spotted sample (4 μg), the difference in the FID response between high and low molecular weight samples is not as pronounced as in the case of S_{II} rods.

These results suggest that an important part of the spotted polymer sample remains spread along the length of the rod. The amount of the unrecovered polymer depends on the polymer molecular weight (is higher for high molecular weight samples) and on the volume of the solvent used in the development (is higher when the solvent volume is small).

TABLE I. THE INFLUENCE OF MOLECULAR WEIGHT
OF POLYSTYRENE SAMPLES ON FID RESPONSE
(ARBITRARY UNITS). MOBILE PHASE: BENZENE

Rods		Polystyrene $\overline{M}_n = 2350$	Polystyrene $\overline{M}_n = 110,000$
	Sample Quantity = 2 µg		
S_{II}	No elution	860	863
	Eluted once	225	113
S_{200}	No elution	854	864
	Eluted once	457	336
	Sample Quantity = 4 µg		
	No elution	1640	1610
	Eluted once	1203	919

All values are mean values of five runs.

In order to verify this assumption, a repeated
development of the same spotted sample on the rods was
carried out with the same solvent to 10 cm, and the
FID response was measured. The results presented in
Figure 5 show that the peaks become narrower with the
increase in number of the repeated developments, and
the value of the FID response increases also.
        Figure 6 shows that by increasing the number of
repeated developments to four times, the entire spotted polymer is recovered, as indicated by the similar
FID response for eluted and uneluted sample. The data
presented in Figure 7 demonstrate that the thinner the
silica layer on the chromarods, the less is the volume
of the solvent available for development, and less
amount of the spotted sample can migrate with the

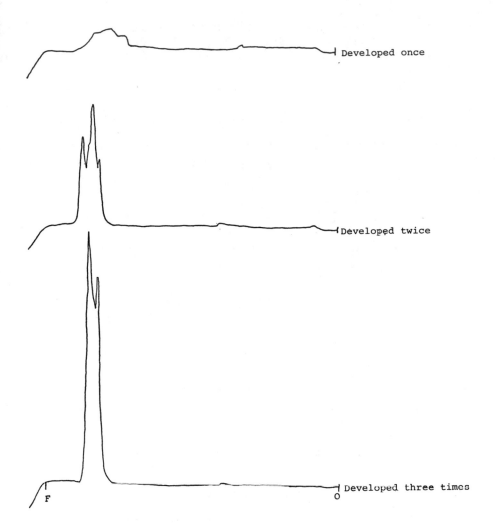

Figure 5. Multiple development with benzene for the same PS sample on the S_{II} chromarods.

solvent front even after three times development of the rod. This tailing effect can be understood by taking into account the fact that the first step in the elution process is the dissolution of the polymer

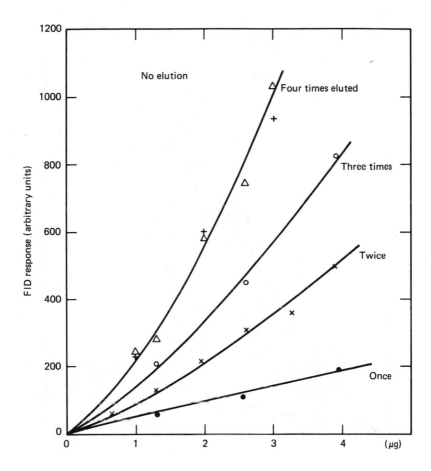

Figure 6. FID response for a PS sample ($\overline{M}_n$ = 200,000) using multiple developments with benzene on chromarods S_{II}.

sample, which is a kinetic phenomenon. The higher the solvent volume that can fill up the pores of the rod where the polymer sample is spotted, the lower the polymer concentration and the easier its dissolution. Due to the intermolecular forces between the polymer chains, not all the polymer chains can dissolve at once. The lower the molecular weight of the polymer,

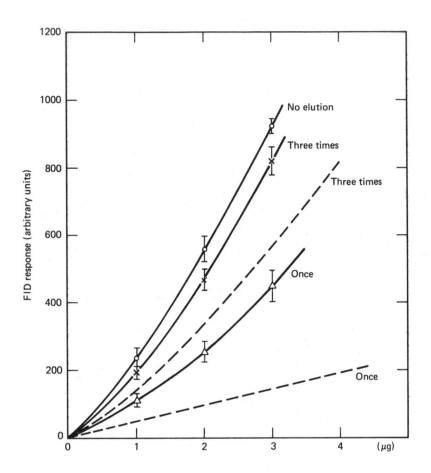

Figure 7. FID response for a PS sample ($\overline{M}_n$ = 200,000) using multiple development with benzene on chromarods S_{200} (———) and S_{II} (-----).

the easier it will dissolve. When the polymer/solvent ratio is low, the dissolution rate increases and at the same time the viscosity of the resulting polymer solution decreases. These facts can explain why in the case of S_{200} rods the amount of the polymer migrating with the solvent front is larger than in case of S_{II} rods. The difference between the two types of

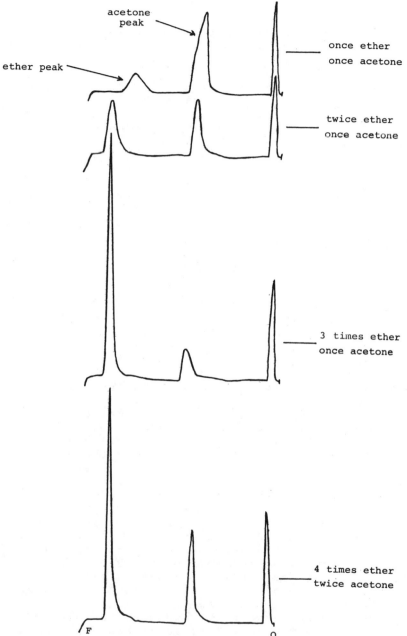

Figure 8. The effect of repeated development on separation of the chromatograms in the two-stage separation of SMB copolymer with diethyl ether for 10 cm followed by development with acetone for 5 cm.

rods becomes more pronounced with an increase in the
amount of the spotted sample (Figure 7). For the same
type of chromarods, each rod has its intrinsic characteristics, and the separation of the same polymer
will be different on each of them. The multiple development reduces these differences and leads to more
reproducible results.

        A very interesting observation was noted during
the analysis of the SMB copolymer sample supplied by
the Iatron Laboratory. The same procedure of two-
stage development was applied to this copolymer sample
followed by scanning of the rods as described in the
Iatron analysis instructions. The separation results
on S_{II} rods were similar to those reported by Iatron.
An attempt was made to find whether or not tailing of
the sample along the length of the rods takes place.
The procedure of multiple development was applied
using diethyl ether to 10 cm, and then, after drying,
the rods were developed with acetone to 5 cm. The FID
response for each experiment is presented in Figure 8,
which shows that the shape and the dimensions of the
peaks are changing completely. With repeated development the ether peak is increasing, while at the same
time the acetone peak is decreasing in value.

        This observation is extremely important from the
point of view of quantitative analysis of the copolymer composition. The data presented in Table II show
that the ratio between the two peaks, representing the
copolymer fractions with different chemical compositions, is dependent on the elution conditions. If
development is carried out several times with ether,
the copolymer fraction that migrates with the ether
solvent is completely recovered from the length of the
rod and can be identified only under this peak. If
the development is done only once or twice with ether,
a large amount of the copolymer fraction corresponding
to this peak remains spread along the rod, and it is
eluted during the second-stage development with acetone. Consequently, the acetone peak reflects not
only the copolymer fraction that migrates with acetone,
but also a part of the copolymer fraction soluble in
ether and still unrecovered from the rod. Under this
condition the ratio of these two peaks will not represent the real ratio between the two copolymer fractions with different chemical compositions. As shown
by the data in Table II, in order to find the real

TABLE II. THE INFLUENCE OF SUCCESSIVE DEVELOPMENTS
ON THE FID RESPONSE (ARBITRARY UNITS) FOR THE
SMB COPOLYMER ON S_{II} CHROMARODS[a]

Development	Total Area	Ether Peak	Acetone Peak	Starting Peak
Once ether Once acetone	1146[b]	194	670	282
Twice ether Once acetone	1120	434	390	296
Three times ether Once acetone	1526	908	334	284
Four times ether Once acetone	1537	942	288	307
Four times ether Twice acetone	1694	928	470	296
No elution	1960			

[a] Spotted amount 10 μg.

[b] All values are mean values of five runs.

ratio between the two peaks corresponding to these two
copolymer fractions with different compositions it is
necessary to develop the rods with each solvent three
or four times in order to be sure that all the copolymer fraction is eluted with the corresponding solvent
front.   It should be stressed that accurate quantitative determinations can be done only in the case of
polymers that can be fractionated with different specific solvents for each polymer fraction.
        The data presented in Table III show the impor-
tance of the thickness of the silica coating on the
rods in improving the quantitative separation of the

TABLE III. THE INFLUENCE OF THE THICKNESS
OF THE SILICA LAYER OF THE ROD
ON THE SMB COPOLYMER FRACTIONATION

Rods	Area of Ether Peak / Area of Acetone Peak				
Type	No. 1	2	3	4	5
S_{II}	0.524	0.565	0.473	0.528	0.555
S_{200}	1.55	0.766	0.979	0.620	0.800

copolymer by TLC-FID. The same copolymer sample was
fractionated on S_{II} and S_{200} rods. The ratio between
the peaks corresponding to the ether-soluble fraction
relative to the acetone fraction are different depending on which rod was used during the separation.
Using the S_{II} rods the amount of the copolymer fraction that migrates with the ether solvent is smaller
than that of the acetone fraction. For S_{200} rods the
ratio between these two fractions changes, increasing
in favor of the ether fraction.

Thus, the experimental data presented on copolymer separation demonstrate that thicker silica layers
on the rods give better results in polymer separations.
However, with both types of rods, using only one
development with each solvent is not sufficient to
obtain the real values of the copolymer composition.

In conclusion, the TLC-FID separation technique
can be successfully applied to quantitative analysis
of polymers only if the polymer can be fractionated
using different solvents for each fraction. It is not
necessary to use a gradient solvent concentration technique for polymer fractionation.

REFERENCES

1. H. Inagaki, H. Matsuda, and F. Kamiyama, Macromolecules 1, 520 (1968).
2. B. G. Belenki and E. S. Gankina, J. Chromatogr. 53,
   3 (1970).

3. J. L. White, D. G. Salladay, D. O. Quinsenberry,
    and D. L. McLean, J. Appl. Polymer Sci. 16, 2811
    (1972).
4. T. Kotaka and J. L. White, Macromolecules 7, 106
    (1974).
5. G. Glockner and D. Kahle, Plaste u. Kautschuk 23,
    338 (1976); Ibid., 23, 577 (1976).
6. F. Kamiyama, H. Matsuda, and H. Inagaki, Makromo-
    lec. Chem. 125, 286 (1969).
7. T. Kotaka, T. Uda, T. Fukuda-Tanaka, and H. Ina-
    gaki, Makromolec. Chem. 176, 1273 (1975).
8. H. Inagaki, T. Kotaka, and T. I. Min, Pure and
    Applied Chem. 46, 61 (1976).
9. F. Kamiyama, H. Inagaki, and T. Kotaka, Polymer J.
    3, 470 (1972).
10. N. Donkai, T. Miyamoto, and H. Inagaki, Polymer J.
    7, 577 (1975).
11. H. Inagaki, T. Miyamoto, and F. Kamiyama, J. Poly-
    mer Sci., Pt. B, 7, 329 (1969).
12. T. Miyamoto and H. Inagaki, Macromolecules 2, 554
    (1969).
13. H. Inagaki and F. Kamiyama, Macromolecules 6, 197
    (1973).
14. N. Donkai, N. Murayama, T. Miyamoto, and H. Ina-
    gaki, Makromolec. Chem. 175, 187 (1974).
15. H. Inagaki, F. Kamiyama, and T. Yagi, Macromole-
    cules 4, 133 (1971).
16. F. Kamiyama, H. Matsuda, and H. Inagaki, Polymer J.
    1, 510 (1970).
17. E. P. Otocka and M. Y. Hellman, Macromolecules 3,
    362 (1970).
18. E. P. Otocka, Macromolecules 3, 691 (1970).
19. E. P. Otocka, M. Y. Hellman, and P. M. Muglio,
    Macromolecules 5, 227 (1972).
20. F. Horii, Y. Ikada, and I. Sakurada, J. Polymer
    Sci., Chem. Ed., 13, 755 (1975).
21. T. Taga and H. Inagaki, Angew. Makromol. Chem. 33,
    129 (1973).
22. H. Inagaki, Adv. Polymer Sci., Molecular Proper-
    ties, 24, 189 (1977).
23. E. P. Otaka, ACS Polymer Preprints, 12, 645 (1971).
24. T. I. Min, T. Miyamoto, and H. Inagaki, Rubber
    Chem. Techn. 50, 63 (1977).
25. T. I. Min, Polymer (Korea) 2, 146 (1978).

CHAPTER 19

Lipids of Human Milk

Joel Bitman
David L. Wood
Nitin R. Mehta
Paul Hamosh
Margit Hamosh

INTRODUCTION

A study has been made comparing the lipid composition
of the milk of mothers who deliver prematurely with
the milk of mothers who deliver at term.   Determination of the lipids in human milk involves a number of
methodological problems that differ considerably from
those involved in determining the lipids of bovine
milk.   The procedures described in this chapter represent continuing efforts to provide better methods of
investigating the lipids of human milk.   These methods
must be applicable to small milk volumes, because
breast milk is available only in very limited amounts.
Moreover, the methods must ensure stability of the
milk lipids during handling, storage, and work-up.
Lipolysis by lipases present in milk must be avoided
to maintain the structures of the secreted lipid components.

EXPERIMENTAL

Representative Sampling: Protocol for Breast
Milk Samples

The amount of human milk available for research is
extremely limited.   Human milk is food for the neonate.
Consequently, only very small amounts can be diverted
to non-nutritional purposes, in contrast to the
bovine, where extremely large volumes are available.
The methods described for complete lipid analysis were
developed for a total volume of 3.5 to 6.0 ml of human
breast milk, sample sizes that were usually available.
      In addition to availability, a major considera-
tion is representative sampling. Hytten (1), by taking consecutive samples during nursing, showed that
initial samples were watery and low in fat, on contrast to those taken at the end of nursing, which had
a very high fat content. Hall (2) recently confirmed
this, finding a threefold increase in total lipid content during the course of nursing.   Thus, in order to
obtain a human milk sample that is representative, it
is important to take an aliquot of the entire contents
from one breast.
      Our protocol consisted of collecting the entire
contents of milk from one breast (at the 9-10 A.M.
feeding) with the Egnell mechanical breast pump
(Egnell Inc., 412 Park Avenue, Cary, Illinois). Milk
was collected from mothers of 18 very premature (VPT,
26-30 weeks gestation age), 28 premature (PT, 31-36
weeks gestation), and six term (T, 37-40 weeks gestation) infants. Samples were taken from an entire
breast collection on postpartum day 3 (colostrum), 7,
21, 42, and 84.
      Samples of milk collected in the hospital were
kept on ice, transferred to the laboratory within 12
hr of collection, and then frozen at -20° C. Samples
collected at home were placed in the freezer compart-
ment of a refrigerator, then brought to the hospital
in Dry Ice and frozen at -20° C. The frozen samples
were thawed at room temperature and samples taken for
either lipid or lipase analysis. The aliquots taken
for lipid analysis were immediately refrozen at -20° C.
At the time of work-up for lipid analysis, the samples
were thawed in cold water, heated rapidly to 80° C,
and held at 80° C for 1.5 min to inactivate the

lipases, since freezing and thawing is known to activate lipases in milk (3).

Gravimetric Fat Determination and Total Fatty Acid Composition

A lipid extract was prepared by homogenizing 1 ml of
lipase-inactivated milk with 18 ml of chloroform-
methanol (2:1, v/v, containing 0.01 percent butylated
hydroxytoluene, BHT, to inhibit lipid oxidation)
according to the method of Folch et al. (4). After
adding 6 ml of 7 percent aqueous sodium chloride, the
phases were separated by centrifugation. After
removal of the aqueous phase, 10 ml of the lower
chloroform phase were collected. Duplicate 1-ml aliquots of the chloroform extract were transferred to
weighed tubes, and the solvent was evaporated under
nitrogen to determine total lipids gravimetrically.
     The total lipid residues were used for the deter-
mination of total fatty acids. After evaporation of
the solvent under nitrogen from a 1 ml aliquot of the
chloroform extract, methyl esters were prepared by
adding 1 ml of the complex methanolic boron trifluo-
ride mixture (5) and heating for 45 min. After adding
1 ml of water, the fatty acid methyl esters were
extracted with 2 ml of hexane. The hexane extracts
were placed in small vials for analysis by gas liquid
chromatography (GLC).

Neutral Lipid Classes

Neutral lipid classes were determined by quantitative
densitometry in situ (6) following separation by two-
stage thin layer chromatography (7). Duplicate 10 µl
aliquots of the chloroform were placed on a 20 x 20 cm
Merck silica gel 60 plate (250 µm layer), which had
been scored to produce 20 individual lanes. The chromatogram was developed first in a saturated tank containing 75 ml of chloroform-methanol-ethanol-acetic
acid (98:2:1:0.1 v/v) until the solvent front advanced
17 cm. After drying under nitrogen in a chamber for
5 min, the plate was developed a second time in a saturated tank containing 75 ml of hexane-ethyl ether-
acetic acid (94:6:0.2 v/v) until the solvent front
reached the top of the plate. The solvent was evaporated from the plate under nitrogen. The plate was

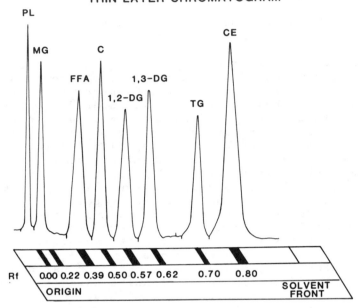

Figure 1. Quantitative densitometry of neutral lipids.
Thin layer chromatogram.

dipped for 3 sec in 10 percent cupric sulfate in 8 percent phosphoric acid (8) and charred by temperature
programmed heating in a gas chromatography oven from
30 to 180° C at 10° C/min.   The two-stage chromatography results in the following separations (Figure 1),
in ascending order from the origin: phospholipids
(R_f = 0.00), monoglycerides (0.22), free fatty acids
(0.39), cholesterol (0.50), 1,2-diglycerides (0.57),
1,3-diglycerides (0.62), triglycerides (0.70), and
cholesteryl esters (0.80).

      Separation of Phospholipid and Neutral Lipid
      Fractions

Fractionation of the total milk lipids into a fraction
containing the neutral lipids and one containing the

phospholipids is useful and necessary for several
reasons. Since the phospholipids are present in
extremely low concentration of 0.5 to 1.0 percent of
total lipid as compared to ca. 97-99 percent triglycerides, it is necessary to concentrate and isolate the
phospholipids to determine their fatty acid components.
Separation of the PL from NL also improves the physical separation of the PL classes on the TLC plate.
Lipid overloading of the TLC plates is avoided when a
fraction containing only the PL is placed on either a
preparative or analytical TLC plate.

Separation of the polar phospholipids from the
neutral lipids has been achieved by column chromatography on silicic acid and Florisil (9). These column
procedures are time-consuming, and careful standardiza-
tion of the adsorbents is necessary to achieve reproducible separations. A separation of PL from NL has
been developed using the commercially available, pre-
packed, disposable silica gel Sep-Pak cartridges of
Waters Associates, Milford, Massachusetts.

Recovery experiments were conducted with a lipid
standard containing 100 mg of triolein and 1 mg each
of sphingomyelin, phosphatidyl choline, phosphatidyl
serine, phosphatidyl inositol, and phosphatidyl ethanolamine to approximate a sample milk lipid in composition. This standard was dissolved in 1 ml of hexane
and placed on a Sep-Pak silica column that had been
previously washed with 20 ml of chloroform. The neutral lipids were eluted by forcing 40 ml of hexane-
ethyl ether (1:1 v/v) through the column with a
syringe. None of the phospholipids were eluted in
this fraction, as demonstrated by quantitative thin
layer chromatography. Two washes of 20 ml each with
more polar solvents, methanol (MeOH) and $CHCl_3$-MeOH-
H_2O (3:5:2 v/v), removed 80 and 20 percent, respectively, of the individual PL components, again measured by quantitative TLC densitometry. The standard
procedure is given below.

A 2.5 to 5.0 ml sample of lipase-inactivated milk
was extracted with chloroform-methanol (2:1 v/v) containing 0.01 percent BHT. The chloroform residue was
dissolved in 2 ml of hexane and transferred to a silica Sep-Pak cartridge previously washed with 20 ml of
chloroform. The residue tube was rinsed with 1 ml of
hexane, and the rinsings were also transferred to the
Sep-Pak column. After the hexane had dripped through

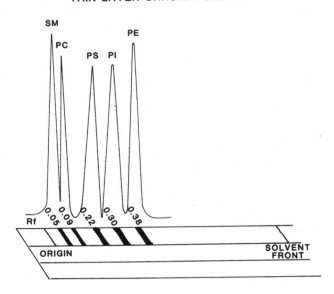

Figure 2. Quantitative densitometry of phospholipids.
Thin layer chromatogram.

by gravity, the nonpolar lipids were eluted by pumping
4 x 10 ml aliquots of hexane-ethyl ether (1:1 v/v)
through the cartridge with a syringe. The eluate contains the neutral lipid fraction (NL). The phospholipids (PL) were then eluted by pumping 2 x 10 ml of
methanol and 2 x 10 ml of chloroform-methanol-water
(3:5:2 v/v) through the cartridge. After evaporation
of solvents under nitrogen, the NL and PL fractions
were dissolved in 1 ml of chloroform containing 0.01
percent BHT to inhibit lipid oxidation.

        Phospholipid Class Separation

Duplicate 10-µl aliquots of the PL fraction from the
Sep-Pak were placed onto the preadsorbent zone of a
20 x 20 cm Merck silica gel 60 plate containing 10 mm
lanes. The plate was developed in a saturated tank

containing 75 ml of chloroform-propanol-ethyl acetate-
methanol-0.25 percent aqueous potassium chloride
(23:25:25:15:9 v/v) until the solvent front reached
the top of the plate, ca. 1.75 hr (10). Solvents were
evaporated from the plate in a chamber under nitrogen
for 30 min. The phospholipids were visualized by
dipping for 3 sec in 10 percent cupric sulfate in 8
percent phosphoric acid and charred by temperature programmed heating in a gas chromatography oven from
30-180° C at 10° C/min. The phospholipids were separated (Figure 2 in the following order from the origin:
sphingomyelin (R_f = 0.05), phosphatidyl choline (0.09),
phosphatidyl serine (0.22), phosphatidyl inositol
(0.30), and phosphatidyl ethanolamine (0.38).

Cholesteryl Ester Fatty Acid Composition

The NL fraction was streaked as a band across 15 cm of
the 20-cm width of the preadsorbent zone of a 20 x 20
cm preparative thin layer plate (1000 μm silica gel
layer, Whatman PLK5). A standard mixture of neutral
lipids was streaked onto the remaining preadsorbent
area. The chromatogram was developed in a tank (saturated conditions) containing 200 ml of hexane-ethyl
ether (95:5 v/v) until the solvent reached the top of
the plate. After drying under nitrogen, the cholesteryl ester fraction was isolated by visualization of
the standard lane after spraying with 0.05 percent
ethanolic Rhodamine B. The cholesteryl ester fraction
was then scraped from the TLC plate and the silica gel
scrapings placed in a 15 ml glass culture tube. The
cholesteryl esters were transmethylated using the
complex BF_3 mixture of Morrison and Smith (5). The
fatty acid methyl esters were extracted into hexane
and transferred to micro GLC vials. After evaporation
of solvent, the residue was redissolved in 50 μl of
hexane for GLC analysis.

Phospholipid Class Fatty Acid Composition

The PL fraction was streaked as a band across 9 cm of
the 20 cm width of the preadsorbent zone of a 20 x 20
cm Merck silica gel 60 analytical glass. The plate
was placed at an angle in a short glass developing
tank (10 cm high) so that the TLC plate extended outside the chamber. The chromatogram was developed

continuously for 3 hr using the solvent system of
Sherma and Touchstone (10), chloroform-n-propanol-
ethyl acetate-methanol-0.25 percent aqueous potassium
chloride (25:25:25:13:9 v/v). A 2-cm strip at each
side edge of the glass TLC plate had been prescored
with a glass cutter on the underside of the glass
plate before sample application. After development,
these glass strips were snapped off and charred with
the cupric sulfate visualization reagent. The rest of
the plate was sprayed with 0.05 percent ethanolic
Rhodamine B to aid in locating the PL classes. After
visualization under UV light, the silica gel bands
were scraped into tubes and methyl esters of the phospholipid fatty acids formed by addition of 1 ml of
percent BF_3 in methanol. The methyl esters were
extracted into hexane and transferred to micro GLC
vials. After evaporation of solvent, the residue was
redissolved in 50 µl of hexane for GLC analysis.

Quantitative Densitometry in Situ

The charred TLC plates containing the neutral lipids
and phospholipids were quantitatively determined by
spectrophotodensitometry in situ using the Shimadzu
CS-910 Dual Wavelength TLC Scanner (Shimadzu Scientific Instruments, Inc., Columbia, Maryland). Plates
were scanned in the linear mode at 350 nm. Optical
densities of the charred spots were compared to neutral lipid or phospholipid standards on each TLC plate.
Peaks were integrated with a Hewlett-Packard 5840 GC
terminal, which was interfaced to a DEC PDP-11 computer (Digital Equipment Corporation, Marlboro, Massachusetts) for postrun analysis of peak data. Data
from standard plates (duplicate 0.2, 0.5, 1, 2, 5, 10,
20 µg lanes) were used to compute second-order poly-
nomial standard curves. Sample concentrations were
determined by computer fitting of the sample data to
the standard curves.

Gas Liquid Chromatography of Fatty Acid Methyl
Esters

Fatty acid methyl esters were analyzed by temperature-
programmed gas liquid chromatography using a 6 ft x
2 mm i.d. glass column packed with 10 percent SP-2340
(Supelco Inc., Bennefonte, Pennsylvania) on Chromosorb

TABLE I. FAT CONTENT OF BREAST MILK. COMPARISON OF GRAVIMETRIC AND TLC DENSITOMETRY RESULTS

Method	Percent Fat at Lactation Day				
	3	7	21	42	84
Gravimetric	2.04[c]	2.89[b]	3.45[ab]	3.19[ab]	4.87[a]
TLC	1.73[b]	2.76[a]	3.43[a]	3.44[a]	3.78[a]
N	39	41	25	18	8

Means with the same letter within a row are not significantly different at the 0.05 probability level.

WAW (100/120 mesh) in a model 5840 Hewlett Packard gas chromatograph. The temperature of the column was held at 50° C for 2 min, then programmed at 8° C/min to 200° C (He flow rate, 30 ml/min; dual FID detectors, detector temperatore 250° C; injection port temperature, 225° C). Peak areas were measured with an electronic integrator. Peak identities and quantitative accuracy were determined from known standards for each fatty acid.

Statistical Procedures

All data were analyzed as two-way designs (group × time) by the General Linear Models Procedure of the Statistical Analysis System (11). When significant differences were present, main effect means were compared by Duncan's new multiple range test. All means presented in tables are least square means since an unbalanced design resulted because of the difficulty in obtaining breast milk samples from the mothers. Experimental parameter pairs were compared by calculation of correlation coefficients based upon simple regression analysis.

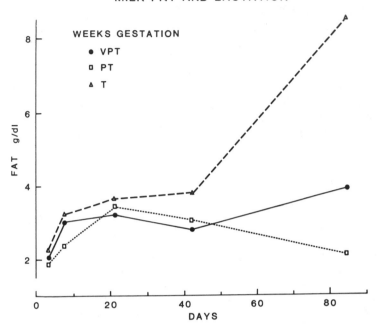

Figure 3. Fat content of human milk during lactation.
Milk was from mothers delivering at 26-30 weeks of gestation (VPT), 31-36 weeks (PT) and 37-40 weeks (T).

RESULTS AND DISCUSSION

Fat Content of Breast Milk.   Comparison of Gravi-
metric and TLC Densitometry

Table I shows the fat content of all samples grouped
by time periods. The fat content of colostrum (3
days) was significantly lower than the fat content of
transitional (7 days) and mature milk (days 21, 42,
and 84). Similar values were obtained from the gravi-
metric determination based on weighing about 30 mg of
lipid residue from 1 ml of milk and from TLC densitometry, in which about 25 µg, 1/1000th of the total

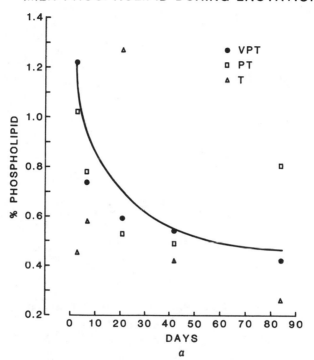

Figure 4 (A). Phospholipid content of human milk during lactation. Milk was from mothers delivering at
26-30 weeks of gestation (VPT), 31-36 weeks (PT), and
37-40 weeks (T). Points are means of values for
groups that contained unequal numbers of samples.
Lines shown are regression lines: percent phospholipid = $1.25 \ x^{-0.23}$; cholesterol (mg/dl) =
$17.54x^{-0.12}$, where x = days.

lipid, were spotted, separated by TLC, and determined
by quantitative densitometry. For all 131 samples of
Table I, total fat content, determined gravimetrically
over all groups and time periods, was 2.78 ± 0.15

MILK CHOLESTEROL DURING LACTATION

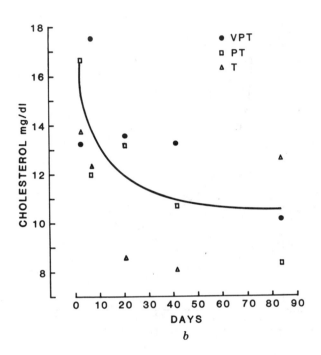

Figure 4 (B). Cholesterol content of human milk during lactation. See Figure 4 (A) for details.

percent. Estimated by TLC densitometry, fat content was 2.70 ± 0.14 percent. The two estimates were highly correlated ($r^2 = 0.82$, p = 0.001).

The fat content data for the three maternal groups during lactation are shown in Figure 3. During the first 21 days of lactation, fat content increased significantly in all groups (regression equation, Y = 2.31 + 0.03x, where Y = fat concentration in g/dl, x = days). Fat content increased about 40 percent by 1 week. Fat content for all groups was constant from 21 to 42 days (Figure 3). Fat percentage at 84 days was not significantly different from these values.

Phospholipid and Cholesterol Content

As the fat content of milk changed during lactation,
increasing in all groups from a low level in colostrum
to a higher level which was maintained throughout lactation, cholesterol and phospholipids exhibited a continuous decline (Figure 4). Both of these components
are structural lipids of the milk fat globule membrane.
Figure 4(A) shows the phospholipid data. Phospholipids are present in colostrum at a concentration over
1 percent but constitute about 0.5 percent of total
lipids in human milk during most of lactation. Figure
4(B) shows cholesterol data during lactation. The
cholesterol content of colostrum is high and decreases
to lower levels in transitional and mature milk. During most of lactation the level of cholesterol was
about 12 mg/dl.

Cholesterol and phospholipids exist in milk as
components of the milk fat globule membrane. The
decline in phospholipids and cholesterol best fit a
power curve ($Y = ax^b$). The curves were: percent phospholipid = $1.25x^{-0.23}$ ($r^2 = 0.95$); cholesterol (mg/dl)
= $17.54x^{-0.12}$ ($r^2 = 0.99$), where x = days. The
decline in both of these components suggests that
there is less membrane as lactation proceeds. Since
total milk lipids increase, these data suggest that
either the size of the milk fat globule increases or
if more fat droplets are secreted, that the membrane
may be thinner as lactation proceeds.

Lipid Classes of Human Milk

Accurate classification of the lipids of human milk
requires proper sampling, storage, and the application
of satisfactory analytical techniques. The lipid
class composition of human milk from mothers of pre-
term and term infants at 21 days of lactation is shown
in Table II, column A. The relatively large percentage of MGs and FFAs were undoubtedly artifacts produced by lipolysis. These samples had been stored at
-20° C for several months but also had been frozen and
thawed twice before actual lipid analysis. Human milk
contains two lipases, lipoprotein lipase and bile salt-
stimulated lipase. Freezing and thawing results in an
activation of these enzymes and an increase in lipolytic activity (3). Recalculation of the mean results,

TABLE II. LIPID CLASS COMPOSITION OF HUMAN MILK
(21 DAYS OF LACTATION) AND BOVINE MILK

Class[b]	Percent in Lipid Class[a]		
	Human Milk		Bovine Milk
	A (Stored at -20° C)	B (Corrected)	C
PL	0.66	0.80	1.20
MG	1.84	0.0	0.0
FFA	16.66	0.0	0.0
CHOL	0.36	0.50	0.33
1,2-DG	0.94	0.0	1.03
1,3-DG	0.70	0.0	0.0
TG	78.80	98.70	97.44
N	25	25	10

[a] Means.

[b] PL (phospholipids), MG (monoglycerides), FFA (free
fatty acids), CHOL (cholesterol), 1,2-DG (1,2-
diglycerides), 1,3-DG (1,3-diglycerides), TG (triglycerides).

[c] Specimens were frozen and thawed twice.

[d] Data were recalculated with addition of the FFA, MG,
and DG to the TG fraction.

assuming no hydrolysis of the triglycerides, yielded
the corrected values shown in column B. Values of ca.
98 percent for the triglyceride content of human milk
agree with results of Bracco et al. (12) who published
the only previous data on the class composition of
human milk lipids. The lipid class composition of
bovine milk, taken from 10 Holstein cows in midlacta-
tion and subjected to the same storage at -20° C,
handling, and work-up methodology (repeated freezing
and thawing) in our laboratory, is shown in column C.
The data are in marked contrast to the pattern

TABLE III. COMPARISON OF LIPID CLASS COMPOSITION
OF HUMAN MILK AFTER IMMEDIATE SOLVENT EXTRACTION
WITH EXTRACTION AFTER STORAGE

Class[b]	Percent in Lipid Class[a]		
	Immediate Extraction	-70° C	-20° C
PL	0.81	0.81	0.81
MG	0.0	0.02	0.26
FFA	0.08	0.29	3.62
CHOL	0.34	0.36	0.36
1,2-DG	0.01	0.05	0.61
TG	98.76	98.45	94.34
N	6	6	6

[a] Means.

[b] PL (phospholipids, MG (monoglycerides), FFA (free
fatty acids), CHOL (cholesterol), 1,2-DG (1,2-
diglycerides), TG (triglycerides).

observed in human milk, the absence of free fatty acid
suggesting a much lower activity of endogenous lipases
and a much greater stability of the triglycerides.

An experiment was conducted on several freshly
obtained human breast milk samples in order to examine
whether hydrolysis of triglycerides had occurred in
our samples. The protocol for the experiment was as
follows: a fresh breast milk sample was obtained, one
portion was immediately extracted with chloroform-
methanol (2:1 v/v), another portion was frozen on Dry
Ice and maintained at -70° C until analysis, and a
third aliquot was frozen at -20° C until analyzed.

A comparison of immediate chloroform-methanol
extraction of freshly obtained breast milk with chloroform-methanol extraction of milk samples frozen and
maintained at -70° C or at -20° C is presented in
Table III. Immediate extraction essentially prevented
lipolysis of the triglyceride, and only very minor

amounts of FFA and 1,2-DG were observed. Milk samples
stored at -70° C and -20° C showed progressively
larger amounts of these hydrolysis products. In samples stored at -70° C, the FFA fraction was only ca.
0.3 percent of total lipid, but this increased to 3.6
percent in those stored at -20° C. The results provided support to the conclusion that extensive hydrolysis had occurred in human milk samples shown in Table
II.

Two factors, length of storage time, and freezing
and thawing, may have been responsible for the much
larger proportion of FFA in the -20° C samples shown
in Table II as compared to those of Table III (17 percent vs. 4 percent FFA). Results in Table II were
from milk samples collected and stored at -20° C over
a 6-18 month period and frozen and thawed twice. Samples in the comparative extraction study (Table III)
had been stored at -20° C for 2-5 months and were
frozen and thawed only once immediately prior to lipid
extraction. The relative contribution of these factors to lipolysis cannot be assessed.

The caution stated by Jensen et al. (13) cannot
be overemphasized: "If human milk lipid classes are
to be analyzed, the milk lipids should be extracted
immediately after being obtained or the lipolytic
activity inactivated by prompt heating to prevent the
formation of DGs, MGs, and FFAs." It is apparent that
almost all of the lipid in human milk exists in the
form of triglyceride. In the data reported by Bracco
et al. (12), 98.1 percent of the lipids were present
as TG in fresh, pooled human milk samples from a milk
bank. Within 36 hr after collection, the milk was pas-
teurized by heating at 63° C for 30 min, which probably inactivated the lipases. To preserve the structure and composition of human milk lipids, it is imper-
ative to either extract samples immediately into
organic solvents, freeze, and store at -70° C, or inactivate lipases by heating or addition of lipolytic
enzyme inhibitors. Our results demonstrate that if
human milk samples were extracted immediately with
chloroform-methanol or were stored at -70° C, the composition of the lipids present was maintained, as evi-
denced by stability of the triacylglycerol structure,
the form in which most of the milk lipids exist.

The rise in FFA levels during storage of human
milk at -20° C might be relevant to human milk banking,

especially when the mother's milk is stored fresh-frozen for feeding her own infant at a later time. This is most frequently the case in severe prematurity because very low birth weight infants are usually fed first parenterally, followed by gavage feeding of their own mother's stored milk. Fresh-frozen human milk might have to be stored at temperatures lower than $-20°$ C for prevention of lipolysis.

REFERENCES

1. F. E. Hytten, Brit. Med. J. 1, 175 (1954).
2. B. Hall, Am. J. Clin. Nutr. 32, 304 (1979).
3. D. P. Schwarz and O. W. Parks, in Fundamentals of Dairy Chemistry, B. H. Webb, A. H. Johnson, and J. A. Alford (Eds.), Avi Publishing Co., Inc., Westport, Conn., 1974, p. 220.
4. J. Folch, M. Lees, and G. H. Sloane-Stanley, J. Biol. Chem. 226, 497 (1957).
5. W. R. Morrison and L. M. Smith, J. Lipid Res. 5, 600 (1964).
6. J. Bitman and D. L. Wood, J. Liq. Chromatogr. 4, 1023 (1981).
7. J. Bitman, D. L. Wood, and J. M. Ruth, J. Liq. Chromatogr. 4, 1007 (1981).
8. J. Bitman and D. L. Wood, J. Liq. Chromatogr. 5, 1155 (1982).
9. M. Kates, in Laboratory Techniques in Biochemistry and Molecular Biology, vol. 3, T. S. Work and E. Work (Eds.), Elsevier Publishing Co., New York, 1972, p. 393.
10. J. Sherma and J. C. Touchstone, J. High Resolut. Chromatogr. Chromatogr. Commun. 2, 199 (1979).
11. A. J. Barr, J. H. Goodnight, J. P. Sall, and J. T. Helwig, Statistical Analysis Systems, Statistical Analysis Systems Institute Inc., Raleigh, N. C., 1976.
12. U. Bracco, J. Hidalgo, and H. Bohren, J. Dairy Sci. 55, 165 (1972).
13. R. G. Jensen, M. M. Hagerty, and K. E. McMahon, Am. J. Clin. Nutr. 31, 990 (1978).

CHAPTER 20
Reversed-Phase Thin Layer Chromatography Selection of Cephalosporium Mutants

Hansruedi Felix

INTRODUCTION

Problems with the wetting of RPTLC plates have been
overcome by the introduction of the precoated RPTLC
plates OPTI UP C12 (1). The plates are compatible
with most reversed-phase solvents including 100 percent water. This is of importance not only for choice
of mobile phase, but also in detection, since water-
based visualizing reagents may be used. It also
allows the application of the plates on agar with a
bacterium seed layer for bioautography (2, 3).
    The goal of the following work was to find new
intermediate products in the biosynthesis of cephalosporin C (Figure 1). It can be assumed that all intermediates contain sulfur. That allows the use of
$(^{-35}S-SO_4^{2-}$ as a marker. Products synthesized from the
tripeptide L-α-aminoadipoyl-L-cysteinyl-D-valine are
antibiotically active substances.
    OPTI UP C12 RPTLC plates are useful in several
respects for the selection and analysis of Cephalosporium acremonium mutants blocked in the biosynthesis
of cephalosporin C:
        1. Intermediate products can be characterized bio-
           logically (3). As penicillin N has no UV

279

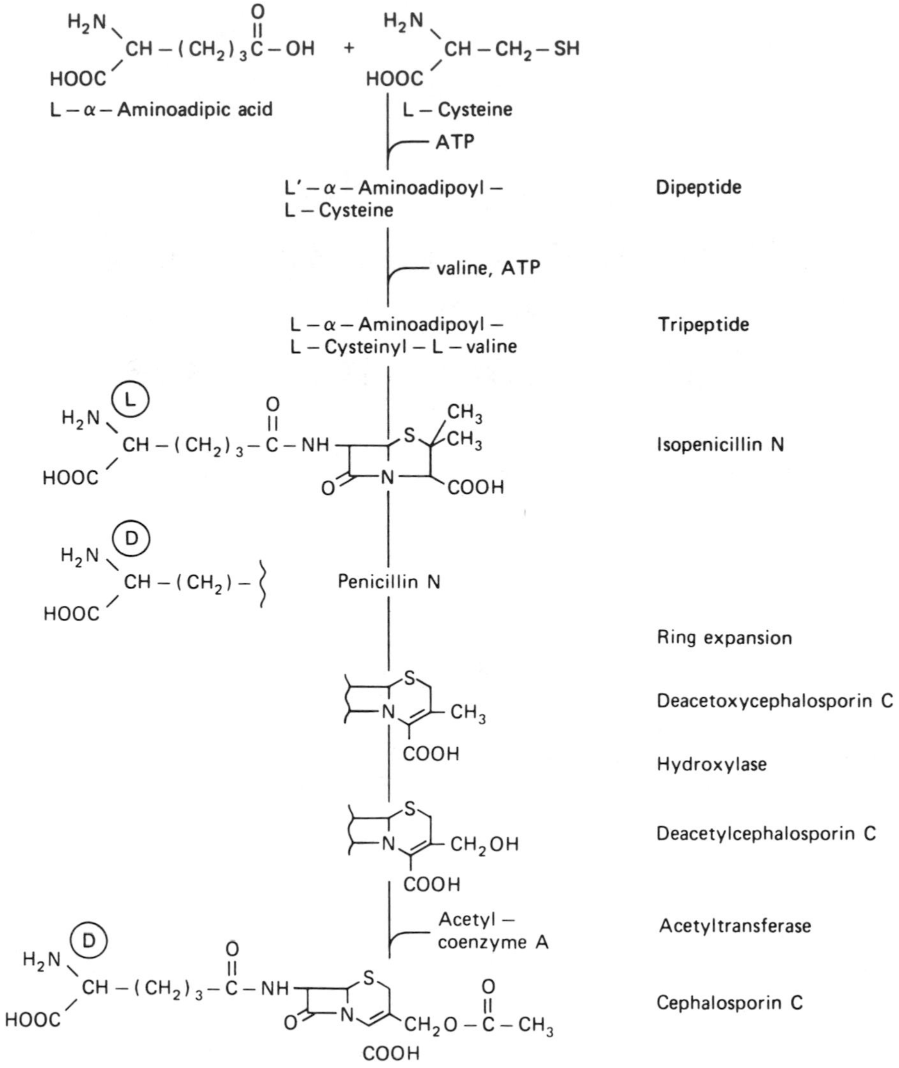

Figure 1. Biosynthesis of cephalosporin C.

active chromophore at 254 nm, it cannot be
detected by the HPLC method used for cephalosporins. For a first screening of the penicillin N amounts, the RPTLC plates were more
suitable than the use of a HPLC system monitored at 200 nm.

2. Radioactively labeled intermediates can be
separated and scanned by scintillation counting (4).

3. Many supposed intermediates and coenzymes are
visible under UV light and after reaction with
ninhydrin (3).

4. By using permeabilized C. acremonium cells,
enzymatic reactions can be followed by applying reaction mixtures directly on the plates
(4-6).

MATERIALS AND METHODS

Microorganisms, culture media and conditions, selection of cephalosporin C-negative mutants, and preparation of enzyme systems have been described (5, 6).

Bioautography

Fifteen μl of a culture filtrate are applied on
10 x 20 cm OPTI UP C12 plates; 100 percent water was
used as the mobile phase. The front reached 15 cm
after about 90 min. Plates were dried in a room temperature air-stream. The RPTLC plates were placed on
Neisseria catarrhalis (or Alcaligenes faecalis or Sarcina lutea) bioautography plates (22 x 15 cm). The
preparation of these agar plates is described elsewhere (3). No air bubbles should be present between
the agar surface and the TLC plate. The TLC plate
together with the agar plate are kept at 4° C for 30
min in order to allow diffusion of possible intermediate products into the inoculated agar. The TLC plate
is then removed and the agar plate incubated at 37° C
for 24 hr. Substances with antimicrobial activity are
visible as clear zones, without the growth of bacteria.
All β-lactams (cephalosporin C, deacetylcephalosporin
C, penicillin N, isopenicillin N) can be detected with
Neisseria catarrhalis, the cephalosporins with Alcaligenes faecalis, and the pencillins with Sarcina lutea.

Even a quantitative determination of the antibiotics
is possible using a standard calibration graph constructed with 2 µl of a 0.25-1.5 g/l solution of the
antibiotics (3).

Radioautography

To 18 ml of a C. acremonium culture (6), 20 µl of
$(S^{-35})-SO_4^{2}$ (5 m Ci/ml) were added after 28 hr of incubation. The incubation continued for 92 hr. The
cells were centrifuged down. Unused $(S^{-35})-SO_4^{2-}$ was
precipitated by $BaCl_2$ (final concentration 0.2M).
Four µl of the supernatant were applied on RPTLC
plates and developed in water (front at 15 cm). The
plates were first screened under UV light at 254 nm
and then sprayed with ninhydrin. For the localization
of S-35-containing species, spots were checked with a
Röntgen film (Kodak XR-5, X-omat, 13 x 18 cm). Film
and plates were held in close contact at -60° C for 7
days. Black spots were marked on the RPTLC plates.
The spots were scraped off. Radioactivity was determined by scintillation counting. A scanner can be
used also, thus omitting the scraping step.

Characterization of the Tripeptide with an
Amino Acid Analyzer

On each of two OPTI UP C12-RPTLC plates (20 x 20 cm),
200 µl of the culture filtrate of C. acremonium
mutant M 269 were applied. The plates were developed
in water-dioxane (98:2 v/v). The band with an R_f
value of 0.26 (dimer tripeptide) was scraped off and
suspended in 20 ml of methanol-water (1:1 v/v). The
solvents were evaporated at 60° C with a rotary evapo-
rator after filtration. Residues were suspended in
100 µl of water. This solution was once again applied
on an RPTLC plate, and the tripeptide band was
finally scraped off. The scraped band, where the tripeptide was assumed to be, was boiled in 6N HCl for 24
hr at 120° C after the extraction procedure (see
above). Water and HCl were evaporated. The residues
were analyzed in an amino acid analyzer (Biotronik
chromatography system LC 6000E). α-Aminoadipic acid
and valine were detectable. Cysteine is destroyed
during hydrolysis.

TABLE I. DISTRIBUTION OF <u>C</u>. <u>ACREMONIUM</u> MUTANTS

Category	Cephalosporins (g/1)	Penicillin N (g/1)	Number of Mutants
1	0	0-0.1	139
2	0-0.1	>0.1	16
3	0.1-1.0	0.1-1.0	31
4	1.3/4.3[a]	2/5	2

Thin Layer Plates

Commercially available precoated OPTI UP C12 RPTLC
plates were used. Plates with or without UV indicator
can be obtained. They are available through ANTEC AG
(4431 Bennwil, Switzerland), Tridom Chemicals (255
Oser Avenue, Hauppauge, New York 11787), Siccap-Emmop
(Marseille, France), and Fluka (Neu-Ulm, GFR).

RESULTS

Depending on the selection method (6), two main types
of Cephalosporium acremonium mutants could be distin-
guished: penicillin N positive/cephalosporin negative
mutants and β-lactam (= penicillins and cephalosporins)
negative mutants. Mutants (188) were checked for
their penicillin N production, final concentrations of
cephalosporins, and for their excretion of S-35-
containing compounds. Amounts of β-lactams greater
than 0.1 g/1 can be determined quantitatively by bioautography (good values between 0.1 and 1 g/1).
Cephalosporins were analyzed additionally by HPLC (6).
These determinations allow the distribution of the 188
mutants into four categories (Table I).
   Mutants of category 1 are nearly completely
blocked in the biosynthesis of all β-lactam antibiotics. Among them mutants excreting peptides or
other intermediates should occur. Analysis of the
culture filtrates on RPTLC plates shows no excretion
of such products in larger amounts than 200 mg/1
(= limit of possible analysis), which is very low

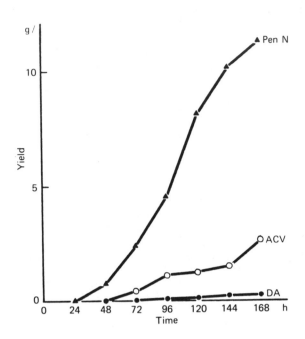

Figure 2. Synthesis of deacetylcephalosporin C, penicillin N, and α-aminoadipoylcysteinylvaline by the
mutant M269. Penicillin N (▲) and α-aminoadipoylcysteinylvaline (o) were determined by ^{35}S-incorpora-
tion with RPTLC; deacetylcephalosporin C (●) or by
HPLC.

compared to the yields in the parent strain (about
10 g/l cephalosporin C). Interesting spots could be
dissolved by the addition of lactamase I (for penicillins) or lactamase P-99 (for cephalosporins). Even
after cell disruption by ultrasonic treatment (7) of
mutants, no intermediates were detectable in the super-
natants using radioautography. Only six out of 139
mutants produced absolutely no penicillin N, the
others always around 80 mg/l.
        Mutants of category 2 produced larger amounts of
penicillin N. Four mutants excreted 5-10 g/l of penicillin N. One mutant was further investigated. A
production curve can be found in Figure 2. Penicillin

N, the tripeptide α-aminoadipoylcysteinylvaline, and
very small amounts of deacetylcephalosporin C were synthesized. The tripeptide was produced toward the end
of the fermentation. However, a disruption of the
cells showed that no tripeptide accumulated inside the
cells. The main products, penicillin N and the tripeptide, were further characterized. Penicillin N was
highly purified by preparative HPLC and is described
in detail elsewhere (6). No isopenicillin N was present. The substance behaving on RPTLC plates as the
dimer tripeptide obtained from Takeda contained S-35
and can be converted with dithiothreitol to the
monomer tripeptide, which changes the R_f value from
0.06 to 0.32. The isolated and purified product shows
after hydrolysis strong signals for the amino acids
α-aminoadipic acid and valine (1:1) on the amino acid
analyzer. Only traces of cysteine are visible, as it
is destroyed during hydrolysis. Signals of other
amino acids, e.g., methionine, cystathionine, homocys-
teine) are not detectable.
     Mutants of category 3 have a reduced β-lactam syn-
thesis. No intermediate products are detectable with
radioautography.
     The two mutants in category 4 are deacetylcephalo-
sporin C producers and were found incidentally during
the mutant selection (6) as A. faecalis is less inhib-
ited by deacetylcephalosporin C than by cephalosporin
C. Radioautography showed that deacetylcephalosporin
C producers excrete comparable amounts of penicillin N
(Table I). Cephalosporin C and deacetoxycephalosporin
C were missing. Interestingly enough the penicillin N
producer M269 excretes only this particular cephalosporin (Figure 2).
     Mutants thus obtained can be further analyzed for
their enzymatic activities. C. acremonium cells
treated with ether are capable of converting deacetoxycephalosporin C and deacetylcephalosporin C to cephalosporin C (5) (Figure 3). Table II shows that the terminal enzymes are present in all of the mutants. The
enzymes can be characterized by omission of substrates
except for the hydroxylase in the parent strain I
18.2.15. This phenomenon is discussed elsewhere (5).
The product cephalosporin C can be destroyed by lactamase P-99 (Figure 3; R_f value in water of cephalosporin
C = 0.32, of lactamase-treated cephalosporin C = 0.79).

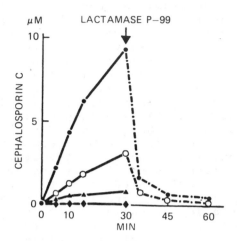

Figure 3. Hydroxylation and acetylation of deacetoxycephalosporin C (DAO) in ether-treated C. acremonium
cells. Ether-treated cells were incubated with DAO,
2-oxo-glutarate, ascorbate, dithiothreitol, Fe^{2+}, and
$1-^{14}C$-acetylcoenzyme A (●)(5). Control experiments
were carried out by omission of DAO (○) or using
boiled ether-treated cells (◆). When using monoxygen-
ase conditions (5), NADH and Mn^{2+} were present (▲).
β-Lactamase P-99 destroyed the product cephalosporin C
(— · — · —). The reaction was followed with RPTLC
monitoring the radioactive product (5).

     The ring expansion enzyme (penicillin N to deace-
toxycephalosporin C = DAO) can be followed by RPTLC
(Figure 4), although not as quantitatively as with
HPLC (6). The product does not occur if lactamase I
is present at the beginning of the reaction. It
destroys the substrate penicillin N. It has no effect
at the end. The product DAO can be destroyed by lactamase P-99 after the ring expansion reaction takes
place. All different mutant types were checked for
the ring expansion activity. The interesting phenomena in this context are discussed elsewhere (6).

TABLE II. COMPARISON OF DIFFERENT C. ACREMONIUM MUTANTS
ANALYZING THE TWO FINAL STEPS
OF THE CEPHALOSPORIN C BIOSYNTHESIS.
THE CULTURES ARE 72 HR OLD.

Strains	Rate of Synthesis of Cephalosporin C[a]				Description of Mutant
	Acetyltransferase		Hydroxylase and Acetyltransferase		
	Addition of DA[b]	-DA[b]	+ DAO[b]	-DA[b]	
I18.215	132	4.4	5.4	4.5	Parent strain
M 269	104	2.9	4.2	0.1	Penicillin N producer
N-2	30	1.1	3.8	0.1	Tripeptide producer (8)
M92	115	3.2	4.0	0.1	No β-lactams

[a] In mg/l hr per g/l RNA.

[b] Substrates (DA = deacetylcephalosporin C, DAO = deacetoxycephalosporin C)

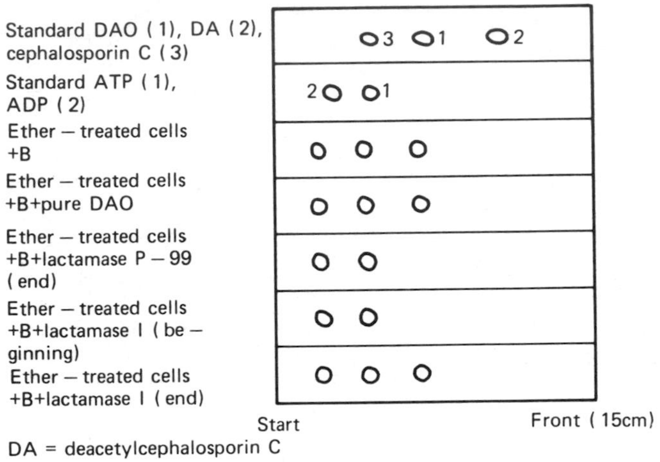

Standard DAO (1), DA (2), cephalosporin C (3)

Standard ATP (1), ADP (2)

Ether — treated cells +B

Ether — treated cells +B+pure DAO

Ether — treated cells +B+lactamase P — 99 (end)

Ether — treated cells +B+lactamase I (beginning)

Ether — treated cells +B+lactamase I (end)

Start   Front (15cm)

DA = deacetylcephalosporin C

Figure 4. Thin layer plate showing conversion of penicillin N to deacetoxycephalosporin C (DAO). The reaction mixture B (2-oxo-glutarate, Fe^{2+}, ascorbate, ATP, penicillin N, for details see Ref. 6) was incubated for 3 hr with ether-treated cells. Ten μl of the tenfold concentrated reaction mixture were applied on RPTLC plates. Spots were detected under 254-nm UV light. Penicillin N has no absorption at 254 nm. DAO was visible down to a concentration of 100 mg/l. Lactamases were added either at the beginning or at the end of the reactions.

        DISCUSSION

RPTLC on OPTI UP C12 plates offers the opportunity to check many different aspects involved in a mutant screening with the same analytical method: chemical and biological characterizations. The method makes it possible to analyze thoroughly a quite large number of mutants. Using a C. acremonium strain growing on sulfate facilitates the experiments. Excess marker S-35-sulfate can be precipitated by Ba^{2+}.
        The cephalosporin and penicillin negative mutants seem to be blocked ahead of the biosynthesis of the peptides or in the biosynthesis of the essential amino acids. As all mutants grew well, they are not strains

unable to utilize sulfate. TLC analysis shows that among the 139 nonproducing mutants, no significant amount of intermediate products occurs. The disruption of 20 mutants released no products with antimicrobial activity or containing S-35 from the cells. The penicillin N producing strain M269 is described elsewhere (6).

Originally it was intended to find peptide-producing mutants. Peptides are only excreted if large amounts of penicillin N or cephalosporins are excreted. Biosynthesis seems to be blocked completely or going at least to penicillin N. There is only one exception known from the literature (8). The penicillin N excreting mutant M269 and nonproducers showed conversion of penicillin N to deacetoxycephalosporin C (6) and significant activity in the two final steps of the biosynthesis. This indicated that the three responsible enzymes are present even if no β-lactams are being synthesized. Thus, these enzymes are not induced by cephalosporin. The cephalosporin C synthesis rate of the two final enzymes were parallel with the in vivo cephalosporin synthesis rates in production strains (e.g., I 18.2.15). In the mutants these enzymes are only present in large amounts at the beginning of the fermentation. Probably the enzymes catalyzing the two final steps are cosynthesized with the ring expansion enzyme. The ring expansion enzyme also is not always present during the fermentation (4, 6).

ACKNOWLEDGMENTS

I am grateful for the many fruitful discussions with Dr. W. Wehrli, Dr. H. J. Treichler, and Professor J. Nueesch (Ciba-Geigy AG, Basel, Switzerland). I wish to thank the company Ciba-Geigy for their financial support while I was preparing my thesis. I thank Dr. R. Binder (M.I.T., Cambridge, Massachusetts) for the critical reading of the manuscript.

REFERENCES

1. M. Faupel and E. von Arx, J. Chromatogr. <u>211</u>, 262
    (1981).
2. E. von Arx and M. Faupel, J. Chromatogr. <u>154</u>, 68
    (1978).
3. M. Faupel, H. R. Felix, and E. von Arx, J. Chroma-
    togr. <u>193</u>, 511 (1980).
4. H. R. Felix, Thesis, Basel, Switzerland, 1980.
5. H. R. Felix, J. Nueesch, and W. Wehrli, FEMS Micro-
    biol. Lett. <u>8</u>, 55 (1980).
6. H. R. Felix, H. H. Peter, and H. J. Treichler, J.
    Antibiot. <u>34</u>, 567 (1981).
7. Y. Sawada, N. A. Solomon, and A. L. Demain, Bio-
    technol. Lett. <u>2</u>, 43 (1980).
8. H. Shirafuji, Y. Fujisawa, M. Kida, T. Kanzaki, and
    M. Yoneda, Agric. Biol. Chem. <u>43</u>, 155 (1979).

CHAPTER 21

Quantification of Phospholipids and Lipids in Saliva and Blood by Thin Layer Chromatography with Densitometry

Carolyn M. Zelop
Leslie H. Koska
Joseph Sherma
Joseph C. Touchstone

INTRODUCTION

In a previous paper (1), a method developed for the densitometric determination of cholesterol in directly applied human saliva using silica gel thin layer plates with a preadsorbent spotting area was reported. Comparisons were made between the cholesterol levels found in a limited number of saliva and fingertip blood samples. In this paper, the method is extended to provide determinations of phospholipids in saliva and blood, and further results are reported for cholesterol and cholesteryl oleate in these samples.

MATERIALS AND METHODS

Whatman 20 x 20 cm LK5D precoated silica gel TLC plates with a preadsorbent spotting area and 19 8-mm-wide scored lanes were used after precleaning by development with chloroform-methanol (1:1 v/v) and air drying.
        The following standard lipids and phospholipids were purchased from Supelco: cholesterol, cholesteryl oleate, lecithin (phosphatidyl choline, egg), spingo myelin (bovine), phosphatidyl ethanolamine (egg),

phosphatidyl glycerol (synthetic), phosphatidyl serine
(bovine), and phosphatidyl inositol (plant). Standard
solutions of the lipids were prepared in chloroform
at concentration levels of 10.0 ng/µl by quantitative
dilution of 1.00 µg/µl stock solutions. Phospholipids
were prepared in chloroform-methanol (1:1 v/v) at 100
ng/µl by dilution of 1.00 µg/µl stock solutions.

Standard and sample solutions were applied by
streaking across the preadsorbent area of each lane as
described earlier (1) using a Drummond 25-µl Diala-
matic microdispenser. This device can be set to apply
any volume from 1 to 25 µl. All samples were Vortex-
mixed for several seconds prior to application and
were dried with cool air from a forced-air gun for 2
min.

For determination of lipids, the chromatogram was
developed with chloroform-ethyl acetate (94:6 v/v).
For phospholipids, ethyl acetate-n-propanol-chloroform-
methanol-0.25 percent aq. KCl (25:25:13:9 v/v) was the
mobile phase (2, 3). TLC plates were developed for a
distance of 15 cm above the bottom edge in rectangular
glass chambers containing a filter paper liner, and
equilibrated with the mobile phase for 15 min before
inserting the plate. Plates containing saliva or
blood samples were predeveloped once with the mobile
phase up to the preadsorbent-silica gel junction,
dried at 120° C for 5 min, and redeveloped in the same
solution (1). Chromatograms were air dried in a fume
hood after development until completely dry (ca. 5
min).

For detection of lipid zones, a dip reagent composed of 30 g of cupric acetate and 80 ml of phosphoric acid per liter of aqueous solution (1) was
used. For phospholipids, both cupric acetate and
cupric sulfate (100 g CuSO$_4$ plus 80 ml of phosphoric
acid per liter) dip solutions were used. Plates were
dipped for 5 sec into the reagent contained in a
Thomas-Mitchell dip tank (A. H. Thomas Co.). The
plate was removed and the excess reagent allowed to
drip back into the tank. Plates were heated in a 160°
C oven for 20 min to produce charred zones.

Scanning was carried out using a Kontes Chromaflex Model K49500 fiber optics densitometer equipped
with a baseline corrector and strip chart recorder.
The single-beam transmission mode was used, with a
scan speed of 4 cm/min and attenuation settings in the

range of 50-200.   The 5-mm light beam was centered on
each lane for scanning.
        Calibration curves were obtained by plotting peak
area × attenuation vs. ng of standards spotted.
Amounts of lipids and phospholipids contained in
applied samples were interpolated from the calibration
plot of standards chromatographed on the same plate.
Concentrations in saliva and blood were calculated
based on the sample aliquot spotted and any correction
factor required because of dilution.
        Saliva was collected upon waking and at other
times of day after rinsing the mouth with water and
waiting for an additional 5 min.   Blood was drawn from
a pricked fingertip into a 1-ml centrifuge tube using
a disposable glass pipet (ca. four free flowing drops).
The sample was centrifuged at 7000 g for 2 min, and
the upper clear, usually colorless or lightly red
serum layer was used.   Dilutions were made quantitatively using a 10-μl Hamilton syringe or Drummond
microdispenser.   All saliva and blood samples were
spotted in duplicate, and results were averaged.

        RESULTS AND DISCUSSION

Initial analyses of saliva and blood showed that the
samples contained the lipids cholesterol and cholesteryl oleate and the phospholipids lecithin, sphingomyelin, and phosphatidyl ethanolamine.   Therefore,
these were the compounds routinely quantified. R_f
values for these compounds in the mobile phases
described above were cholesterol 0.40, cholesteryl
oleate 0.76, lecithin 0.30, sphingomyelin 0.23, and
phosphatidyl ethanolamine 0.58.
        Evaluations with standards of the above compounds
confirmed the earlier conclusion that Whatman scored
K5 preadsorbent TLC plates were superior to HPTLC
plates for densitometric quantification.   This was due
to the greater linearity of the calibration curves and
the ability to spot larger sample aliquots, which offset the lower detection level of the HP plates.   The
5-mm scanning head was centered on each of the 8-mm
lanes prior to scanning; because of the good uniformity of the zone distribution across the width of the
lanes, exactly reproducible centering of the light
beam was not a critical factor in obtaining comparable
scan areas.

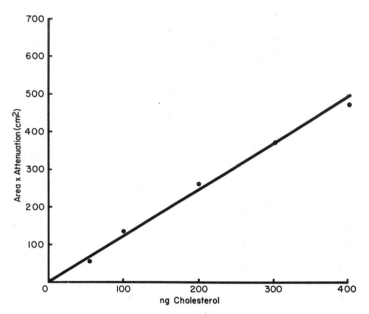

Figure 1. Typical calibration curve for cholesterol on a Whatman LK5D plate resulting from application of 50-400 ng of cholesterol, detection with cupric ace- tate, and scanning with the Kontes fiber optics densi- tometer.

        Calibration curves for the cholesterol and choles- teryl oleate with cupric acetate reagent were linear from 0 to 400 ng and curves for the three phospholip- ids from 0 to 2.5 µg. Figures 1-3 illustrate typical curves for cholesterol, cholesteryl oleate, and the phospholipids. As seen in Figure 1, the calibration curve for cholesterol had a wider linear range than the 260 ng reported earlier (1). This probably was due to the use of a different scanner than in the earlier study. The lower limit of detection and quan- tification for cholesterol and cholesteryl oleate was 50 ng, and for the three phospholipids 300-500 ng. The differences in the slopes for the phospholipid curves (Figure 3) indicate the importance of including standards of each compound to be quantified on the same plate with samples. Linearity coefficients (r) were consistently 0.98 or higher and intercepts within

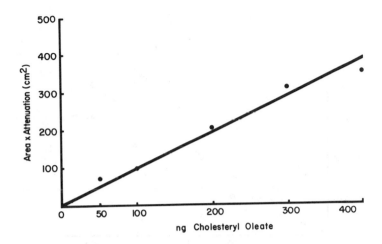

Figure 2. Calibration curve for 50-400 ng of cholesteryl oleate with cupric acid detection reagent.

± 20 of the origin for the calibration curves of all
compounds.
        Cupric acetate reagent was shown to react only
with unsaturated phospholipids in each zone while
cupric sulfate reacts with saturated as well as unsaturated molecules (4). For this reason, a greater
degree of charring was obtained for both the samples
and standards when the cupric sulfate detection
reagent was applied. Saturated and unsaturated phospholipids can be differentiated and quantified using
these two detection reagents in combination, if well-
defined, synthetic standards are available (4, 5).
This was not the case in the present study, therefore
all of the quantitative results below for the three
phospholipids are based on commonly used natural standards that are undefined in terms of degree of saturation and unsaturation. A higher slope was obtained
for the calibration curve for a given phospholipid
when cupric sulfate was used compared to cupric acetate. However, linearity coefficients were usually
lower (ca. 0.96) and intercepts further from zero.
        The level of dilution of the biological samples
and the aliquot volume applied for TLC must be

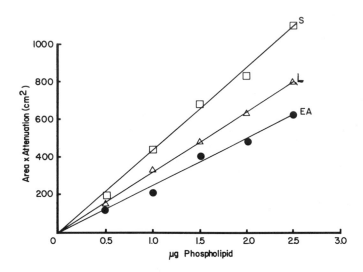

Figure 3. Calibration curves for 500-2500 ng of phosphatidyl ethanolamine (EA), lecithin (L), and sphingomyelin (S) with cupric acetate detection reagent.

adjusted for each sample so that the amount of lipid
or phospholipid to be quantified lies within the linear portion of the respective calibration curve.   As
an example, phospholipid determinations required
spotting of 10-20 µl of undiluted saliva and 5-10 µl
of blood diluted 10:115 (v/v) with distilled water.
(Ten µl was spotted when using cupric acetate detection reagent and 5 µl for cupric sulfate.)   For determination of lipids, saliva was diluted 1:2 (v/v) with
water and 5 µl was spotted.   For blood analysis, 5 µl
was spotted after 10:115 (v/v) dilution for determination of cholesterol and 10:490 (v/v) dilution for
determination of cholesteryl oleate.
     Reproducibility (precision) was evaluated by
spotting four replicate samples across the lanes of a
TLC plate followed by development, detection, and
scanning.   Relative standard deviations for cholesterol were 2.06 percent for 10 µl (100 ng) standards,
4.80 percent for 5 µl of blood (diluted 10:115), and
15.6 percent for 5 µl of saliva (diluted 1:1).   For

phosphatidyl ethanolamine, RSD values were 8.4 percent
for standards (1 µg/µl), 12.9 percent for blood (5 µl/
10:490 dilution), and 9.89 percent for saliva (20 µl/
undiluted). These are reasonable and adequate precision values for microanalysis of biological fluids at
the ng and µg level.

Accuracy of the phospholipid determination was
evaluated by a standard addition procedure, as was
done earlier (1) for cholesterol. A saliva sample was
analyzed, and a value of 400 ng of lecithin was
obtained. This represented 400 ng/20.0 µl of undiluted saliva spotted, or 20.0 ng/µl (2.00 mg/100 ml).
The remaining sample was spiked with an appropriate
volume of a lecithin standard solution to double its
concentration, and the analysis was repeated. After
correcting for the dilution caused by the volume of
standard added, the value of lecithin found was 3.95
mg/100 ml, indicating 98.8 percent recovery of the
spike.

Table I shows typical data for cholesterol levels
determined for a single subject, a healthy female college student, monitored over a period of 19 days in
one month. Morning values ranged from 8.52 to 13.9
mg/100 ml, with an average of 11.21. The morning
value was highest, with levels decreasing until lunch.
Blood values were higher than saliva values, but the
daily and within-day variation trends were generally
similar. Morning blood cholesterol averaged 80.9
mg/ml. Absolute cholesterol values for both saliva
and blood were higher than reported in our earlier
study (1).

The only other neutral lipid found in saliva was
cholesteryl oleate, and Table II shows data for cholesterol and cholesteryl oleate, for the same subject as
in Table I, monitored over a period of about 2 weeks
upon waking and, in a few cases, later in the day.
Except for one sample, the oleate was found in morning
saliva, with values ranging from 8 to 23 mg/100 ml.
Blood values were considerably higher (380-637 mg/
100 ml). The data for cholesterol in Table II are
generally similar in both samples compared to Table I.
It is evident from examination of these tables that
there is no clear, consistent correspondence between
values found for a given compound in the two biological samples or for the two compounds in one of the
samples. That is, increases or decreases in saliva

TABLE I. CHOLESTEROL LEVELS IN mg/100 ml
DETERMINED BY THE PREADSORBENT TLC/DENSITOMETRY PROCEDURE

Date	Saliva				Blood	
	Morning (Before Breakfast)	1 hr After Breakfast	2 hr After Breakfast	3 hr After Breakfast	Morning	1 hr After Breakfast
2-1-82	8.52	5.80	4.96	3.28[a]	--	--
2-2-82	13.9	5.24	4.48	2.52	--	--
2-3-82	13.2	4.0	--	--	--	--
2-15-82	10.0	--	--	--	111	--
2-17-82	9.10	--	--	--	78.8	--
2-18-82	11.3	0	6.0	--	52.5	37.5[b]
2-19-82	12.4	--	--	--	81.3	--

[a] Saliva value after lunch = 1.36 mg/100 ml.

[b] Blood value 2 hr after breakfast = 45.8 mg/100 ml.

TABLE II. MORNING CHOLESTEROL AND CHOLESTERYL OLEATE
          LEVELS IN SALIVA AND BLOOD (mg/100 ml)

Date	Cholesterol		Cholesteryl Oleate	
	Saliva	Blood	Saliva	Blood
3-29-82	13.0	62.5	8.80	380
3-30-82	7.10[a]	65.1[b]	0 [c]	391
3-31-82	10.4	82.2	22.9	467
4-1-82	0.64	99.3	21.9	453
4-12-82[d]	16.3	142	12.2	412
4-13-82	6.22	58.1	14.8	451
4-14-82	13.0	82.9	8.13	532
4-15-82	10.8	151	6.11	637

[a] Before lunch = 5.20; before dinner = 2.31.

[b] Before lunch = 30.8.

[c] Values were also zero before lunch and before dinner.
on this date.

[d] Afternoon values = cholesterol 8.16 (saliva) and 137
(blood); oleate = 7.23 (saliva) and 336 (blood).

values from day to day are not necessarily accompanied by corresponding changes in blood values. Several of the samples in Table II show that the general
trend of decreasing cholesterol values throughout a
day is not always found. The data in Tables I and II
apparently represent normal values and variations in
the values of these compounds in saliva and fingertip
blood, and are probably related to differences in diet
and activities of the subject.

     A single morning saliva sample from a second sub-
ject, a healthy male college student, was analyzed.
Values of 2.56 mg/ml of cholesterol and 7.49 mg/ml of
cholesteryl oleate were obtained, which are significantly lower than the average values for subject 1 but
within the range of values.

     The saliva and blood of subject 1 was also men-
tioned for phospholipid content over the same 7-week

period during which the data in Tables I and II were
collected. Lecithin, sphingomyelin, and phosphatidyl
ethanolamine were found in many, but not all of the
samples, but none of the other phospholipids whose
standards were chromatographed in parallel with the
samples (see Materials and Methods). Table III gives
representative data for phospholipid determinations.
Lecithin in saliva ranged from 2 to 12 mg/100 ml,
sphingomyelin from 0 to 7, and phosphatidyl ethanolamine from 2 to 19. Blood values varied much more
widely, and phosphatidyl ethanolamine was usually
absent. The before-lunch values for two samples
showed both increases and decreases compared to morning levels.

     Lipid and phospholipid saliva levels for a third
subject, also a healthy female college student, were
monitored for over two months. These data are shown
in Table IV for morning, before lunch, and before dinner samples. Morning cholesterol values ranged from
3.35 to 6.01 mg/100 ml and decreased with time during
a single day. Again, cholesteryl oleate was the only
other lipid detected; morning values ranged from 0.4
to 3.5 mg/100 ml, and later samples had both lower and
higher levels. Lecithin, phosphatidyl, ethanolamine,
and sphingomyelin were found in most samples analyzed.
As in the earlier tables, values of zero indicate
amounts below the detection level of the reagent and
not necessarily total absence of a particular compound.
Lecithin ranged from 0 to 3.5 mg/100 ml, phosphatidyl
ethanolamine from 1.0 to 4.6, and sphingomyelin from
0 to 1.6; these values are generally lower than for
subject 1. Phospholipid concentrations usually were
highest in the morning saliva for those days when multiple samples were analyzed.

     As stated above, it has been shown (4) that
cupric sulfate reagent causes charring of both saturated and unsaturated phospholipids, while cupric acetate reacts only with those that are unsaturated.
Reliable quantification of the two types of phospholipids requires the use of synthetic, defined phospholipid standards, which were not available for this
study. However the 4-1-82 saliva and blood samples
(Table II) were also analyzed using cupric sulfate
reagent and the available standards. Values for L, S,
and EA in saliva were 16.3, 18.1, and 15.7 mg/100 ml,
respectively, and in blood 701, 29.4, and 0. All of

TABLE III. MORNING AND BEFORE-LUNCH PHOSPHOLIPID LEVELS
IN SALIVA AND BLOOD (mg/100 ml)

Date	Saliva			Blood		
	L	S	EA	L	S	EA
2-22-82	9.76	1.10	3.32	108	0	113
2-24-82	12.1	1.37	2.50	0	0	0
3-29-82	10.2	0	6.25	306	0	0
3-31-82	7.15	6.58	18.6	198	34.9	0
3-31-82a	7.14	3.80	22.3	249	61.6	48.0
4-1-82	6.83	1.93	17.8	19.7	3.87	6.40
4-12-82	8.24	1.91	9.05	618	49.1	0
4-13-82	2.37	0	10.4	33.1	193	0
4-15-82	7.81	0	--	242	52.3	--
4-15-82a	3.44	4.75	--	205	117	--

L = lecithin.
S = sphingomyelin.
EA = phosphatidyl ethanolamine.
a Values are for samples collected before lunch on these dates.

TABLE IV. LIPID AND PHOSPHOLIPID LEVELS IN SALIVA (mg/100 ml)

Date	Cholesterol	Cholesteryl Oleate	Lecithin	Phosphatidyl Ethanolamine	Sphingo-myelin
9-7-82	5.18	--	--	--	--
9-28-82	6.09	--	--	--	--
10-4-82	3.35	--	--	--	--
10-5-82	5.26	--	--	--	--
10-11-82	5.98	--	--	--	--
10-12-82	6.01	--	--	--	--
10-13-82	4.84	--	--	--	--
10-14-82	3.65	--	--	--	--
10-14-82[a]	1.12	--	--	--	--
10-16-82[a]	6.70	--	--	--	--
10-16-82[a]	2.92	--	--	--	--
10-16-82[b]	2.73	--	--	--	--
10-20-82	--	4.02	--	--	--
10-20-82[a]	--	3.05	--	--	--
10-21-82	--	3.54	--	--	--
10-27-82	--	--	2.64	6.67	0.75
10-28-82	--	--	3.47	6.10	0.84
11-15-82	--	0.43	c	4.29	c
11-16-82	--	0.48	1.22	c	0.53
11-16-82[a]	--	--	0.73	c	1.59
12-4-82	3.75	0.46	1.83	4.01	0.44
12-6-82	4.65	0.67	1.53	4.56	0.48
12-6-82[b]	0.58	1.34	0	0.68	0
12-8-82	3.66	1.02	1.57	3.26	0.59
12-8-82[a]	0.00	0.44	0.49	1.04	0.00

a Sample collected before lunch.   b Sample collected before dinner.
c Compound detected but not quantified.

the saliva values were considerably higher with $CuSO_4$, while the blood values were higher, lower, and the same. These results clearly indicate the importance of the nature of the detection reagent and the dependence of results for phospholipid determinations on the standards used and the detection conditions.

    When saliva samples were diluted with chloroform rather than with water, values for cholesterol were consistently higher. For six different samples, values (mg/100 ml) were 4.08, 3.33; 3.02, 2.43; 3.91, 3.12; 3.31, 2.10; 2.64, 0; 7.88, 5.13, respectively. The significance of these results is not certain, but selective solubility of different forms of cholesterol in the two solvents may be involved. Solvent effects are now being studied further. All samples reported in this paper were diluted with water so as to standardize results.

        SUMMARY AND CONCLUSIONS

A simple, precise, and accurate method has been described for determination of lipid and phospholipid concentrations in blood and saliva based on quantitative preadsorbent TLC with direct sample application. The compounds found and their concentrations are reported for three healthy college students, two of whom were monitored over periods of time. Further work is in progress to evaluate important variables in the analytical procedures, to collect further data, to apply the method to other sample types, and to evaluate the significance of values for diagnostic purposes.

    REFERENCES

1. J. C. Touchstone, C. M. Zelop, and J. Sherma, Proceedings of the 2nd Biennial Symposium on Advances in TLC, Wiley-Interscience, New York, 1982, Chapter 16, p. 219.
2. J. Sherma and J. C. Touchstone, J. High Resolut. Chromatogr. Chromatogr. Commun. 2, 199 (1979).
3. J. C. Touchstone, J. C. Chen, and K. E. Beaver, Lipids 15, 61 (1980).

4. J. C. Touchstone, S. S. Levin, M. F. Dobbins, and
    P. J. Carter, J. High Resolut. Chromatogr. Chroma-
    togr. Commun. 4, 423 (1981).
5. J. C. Touchstone, S. S. Levin, M. F. Dobbins, and
    P. C. Beers, J. Liq. Chromatogr. 6, 179 (1983).

CHAPTER 22

Comparative Analysis of Tocopherols by Thin Layer Chromatography and High-Performance Liquid Chromatography

Tom R. Watkins
Bruce A. Ruggeri
R. J. H. Gray
Richard C. Tomlins

INTRODUCTION

The nutritional and biochemical roles of vitamin E as an antioxidant and free-radical scavenger (1, 2) and membrane lipid stabilizer (3, 4) generally have been recognized. In the past, numerous analytical methods have been employed for the separation and quantitation of the various tocopherol isomers and related compounds.

The oldest and most widely employed method is the Emmerie-Engel oxidometric reaction based on the reduction of $FeCl_3$ to Fe^{2+} by tocopherols, the Fe^{2+} forming a red-colored complex with α,α-dipyridine (5). This complex is measured colorimetrically at 520 nm.

The difficulties and limitations inherent in this method lie in the fact that carotenoids, cholesterol, and vitamin A, and other nonspecific reducing compounds interfere with the colorimetric reaction (6). The unstable color and the variable times for maximum color development are also inherent problems with this method.

In 1961, Tsen developed a modified Emmerie-Engel procedure employing bathophenanthroline, which forms a more stable chromophore with Fe^{2+}. This increases the

sensitivity of the colorimetric reaction two- to
threefold, but still does not eliminate the interfering influences (7).

Alternately, spectrofluorometry is an extremely
sensitive method for assaying free and esterified
tocopherols. The original method of Kofler involved
oxidation of tocopherols with nitric acid to form a
fluorescent phenazine derivative (8). Current methods
involve measuring fluorescence on tocopherol-contain-
ing solvent extracts at 295 and 340 nm, the wavelength
maxima for excitation and emission, respectively.
These methods are preferred over colorimetric procedures due to speed, simplicity, sensitivity, and the
absence of interference from the nonspecific reducing
compounds mentioned previously according to Desai (6).

In our laboratory, we sought to develop an analytical system using thin layer chromatography (TLC) and
high-performance liquid chromatography (HPLC).

Until recently, the use of column chromatography
for the resolution of all the tocopherol isomers had
met with limited success. The advent of high-speed
and high-pressure liquid chromatographic procedures
has led to significant improvements in the separation
and quantitation of these compounds.

A number of workers have demonstrated the advantages of HPLC for tocopherol analysis using normal
phase (9-12) and reversed-phase systems (13, 14). The
advantages include separation of α-, β-, Γ-, and δ-
tocopherols, high specificity and sensitivity, good
reproducibility, and speed and ease of sample application. It should be noted, however, that Hatam and
Kayden (14) were unable to resolve the β- and Γ-
isomers, which chromatographed as a single peak.

TLC is also a widely employed method for tocopherol analysis. Until recently, most thin layer
chromatographic procedures involved scraping of the
resolved tocopherols from silica gel plates and elution with ethanol for subsequent analysis by the colorimetric or spectrofluorometric methods described previously. The procedures described here involve in situ
separation and quantification of tocopherols and
related compounds, circumventing the need for scraping
and elution of the compounds.

METHODS AND MATERIALS

The standard compounds used in this investigation
(dl-α-tocopherol, dl-Γ-tocopherol, dl-δ-tocopherol,
dl-α-tocotrienol, and dl-tocol) were obtained from
Hoffman-LaRoche, Inc. (Nutley, New Jersey). All of
the compounds were dissolved in filtered HPLC grade
methanol (Burdick and Jackson, Muskegon, Michigan)
for subsequent TLC and HPLC analysis.

Thin Layer Chromatography and High-Performance
Liquid Chromatography

Silica gel GF plates (Analtech, Newark, Delaware) were
prewashed in chloroform-methanol (1:1 v/v) and upon
drying, activated at 100° C for 10 min. Samples of
the various compounds were applied (0.2 to 2.4 µg)
under a nitrogen stream in diffuse light to minimize
the danger of peroxidation.

The spotted plates were developed in a mobile
phase consisting of hexane-isopropyl ether (85:15 v/v).
Systems employing 85:15 and 80:20 hexane-ethyl ether
(85:15 v/v) and 80:20 v/v) were attempted, but these
did not give clear separation of α-tocopherol from an
unsaponified triglyceride extract spotted on the
plates. It was our goal to avoid saponification of
our test samples (algal lipids) to reduce any chance
of tocopherol damage. Alternately, a system of ace-
tone-benzene-water (91:30:8 v/v) was attempted, yield-
ing excellent separation of the algal lipids but causing the standards to run off the plates. The hexane-
isopropyl-ether (85:15 v/v) mobile phase proved the
most effective.

The developed plates were air-dried, oven-dried
for 15 min at 100° C, and sprayed with 10 percent
$CuSO_4$ in 8 percent H_3PO_4 followed by charring at 190°
C for exactly 10 min. The resolved compounds were
quantified using a Shimadzu TLC Scanner (Model CS 920)
at 350 nm using a deuterium lamp.

An analytical procedure was developed using HPLC.
The tocopherol standards and related compounds were
analyzed isocratically on a Varian Model 5000 HPLC
equipped with a Vari-Chrom UV/visible spectrophotometer. The column was a reversed-phase Varian MCH 10
C_{18} Micropak column (monomeric). The mobile phase
consisted of methanol-water (95:5 v/v) set at a flow

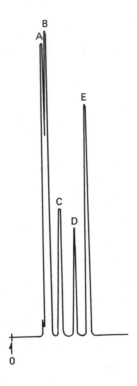

Figure 1. Thin layer chromatogram of tocopherol
isomers and related compounds.   Tocopherol standards:
TLC - silica gel GF.   Solvent: 85:15 hexane/isopropyl
ether, λ = 350 nm, D_2 lamp, 10 mV.   Visualizing agent:
10 percent $CuSO_4$/8 percent H_2PO_4.   Peak identities:
A = dl-tocol (0.64 μg), B = dl-δ-tocopherol (0.4 μg),
C = dl-Γ-tocopherol (0.4 μg), D = dl-α-tocotrienol
(0.4 μg), E = dl-α-tocopherol (0.4 μg). R_f values:
R_{fa} = 0.27, R_{fb} = 0.29, R_{fc} = 0.33, R_{fd} = 0.37, R_{fe} =
0.42

rate of 2.0 ml/min.   The spectrophotometer was set at
a wavelength of 296 nm with a 0.05 absorbance range
and band width of 16 nm.   Samples ranging from 2 to 50
μl were placed on the column.

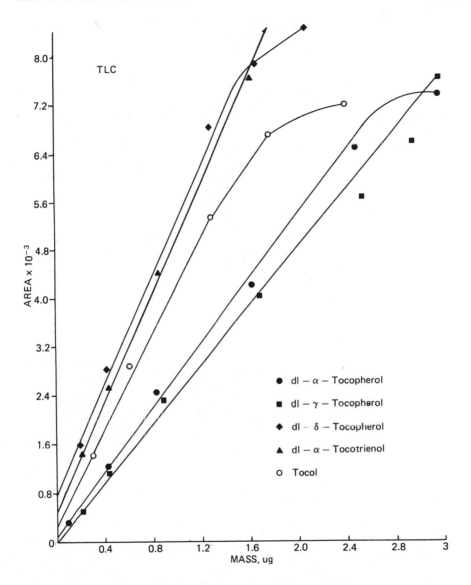

Figure 2. Standard TLS calibration curves for tocopherols and related compounds.

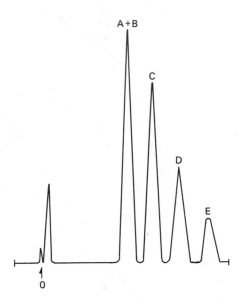

Figure 3. High-performance liquid chromatogram of
tocopherol isomers and related compounds.   Tocopherol
standards: HPLC - HCH 10 C_{18} Micropak Column.   Solvent: 95:5 methanol/water, λ = 296, UV A = 0.05, flow
rate = 2.0 ml/min, P_o = 158 ATM, temperature = 27° C.
Peak identities: A + B = dl-tocol (1.4 µg) + dl-α-
tocotrienol (2.0 µg), C = dl-δ-tocopherol (2.0 µg),
D = α-Γ-tocopherol (2.0 µg), E = dl-α-tocopherol (2.0
µg).   Retention times: A + B = 4.9-5.1 min, C = 6.1
min, D = 7.5 min, E = 9.2 min.

RESULTS OF ANALYSES

Figure 1 illustrates the chromatogram obtained for TLC
analysis of the standard compounds.   Although dl-α-
tocopherol and dl-tocol chromatographed closely, the
scanner was able to quantify distinct peaks, enabling
the resolution and quantification in situ of all of
the compounds of interest.   The lower limit of sensitivity approached 0.2 µg.   At higher concentrations a
tailing effect is observed in agreement with the limitation of the Kubelka-Munk theory.   Reproducibility of
the sample analyses was excellent.   Figure 2

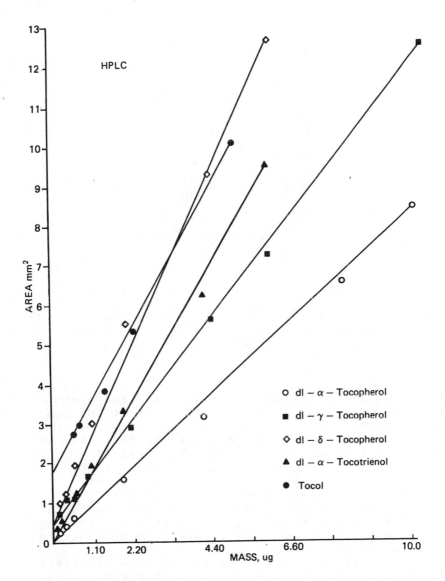

Figure 4. Standard HPLC calibration curves for toco-pherols and related compounds.

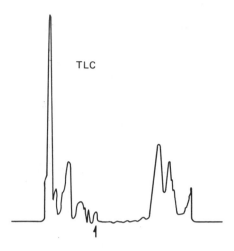

Figure 5 (A). Sample chromatogram (TLC) of lipid
extracts of Euglena gracilis, Strain Z.   TLC - silica
gel GF.   λ = 350 nm, 10 mV range.   Chart speed: 2.0
cm/min.   Sample: E. gracilis, Z, 4 μ   application.
I = initial peak, S = solvent line, α = dl-α-tocopherol, R_f = 0.42.

illustrates the standard curves obtained for each compound and the aforementioned tailing effect for some
of the compounds.
     Recovery studies were conducted on TLC using a
lipid extract from Euglena gracilis, Strain Z to which
500 μg of dl-α-tocopherol was added prior to the
extraction of the cells with HPLC grade acetone
(Burdick and Jackson, Muskegon, Michigan).   Upon
exhaustive extraction, rotary evaporation, resuspen-
sion in HPLC-grade methanol, and filtration, the lipid
extracts were analyzed.   The recovery of α-tocopherol
was 91 percent.
     The results of the HPLC analysis were comparable
to those obtained by TLC.   Figure 3 illustrates the
chromatogram obtained under the stated operating conditions, along with the retention times for each compound.   Dl-tocol and dl-α-tocotrienol co-chromato-
graphed.   This could not be alleviated without impairing resolution of the tocopherol isomers placed on the
column.

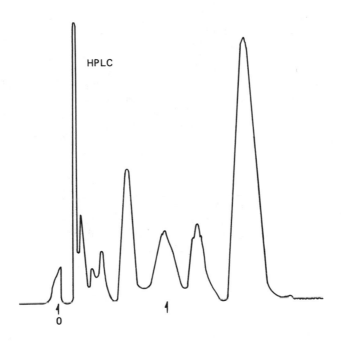

Figure 5 (B). Sample chromatogram (HPLC) of lipid
extracts of Euglena gracilis, Strain Z. HPLC-(RP)C_{18}
Micropak, λ = 296 nm, A = 0.05. Chart speed: 1.0
cm/min. Sample: E. gracilis, Z, 40-μl application.
I = inject peak (solvent, α = dl-α-tocopherol (spiked
sample), R_t = 9.2 min.

        Standard curves were obtained for each compound
individually as shown in Figure 4, and a duplicate
recovery experiment was conducted using the same
spiked extract. A recovery of 91.6 percent was demonstrated, comparable to that obtained for TLC.

        DISCUSSION

The TLC and HPLC systems described appear comparable
in sensitivity, reproducibility, percent recovery, and
ease of application. Figure 5 is a sample chromatogram from TLC and HPLC analysis of the lipid extracts

from a spiked and unspiked sample of cells of E. gracilis, Strain Z for comparison purposes.

All of the compounds analyzed were resolvable by
TLC as shown in Figure 1, while dl-tocol and dl-α-
tocotrienol were not using HPLC (Figure 3). This is
not the major factor, however, in judging the efficiency of the HPLC system. The advantage of one system
over the other lies in analysis time compared to number of samples being analyzed. HPLC appears advantageous for analyzing larger volumes of sample if relatively few in number, along with standards, requiring
10 to 12 min per sample, including a washing between
samples. TLC, although necessitating smaller sample
volumes, is advantageous when analyzing larger numbers
of samples; on a single plate one can spot five or six
standards and six samples in duplicate and run a complete in situ analysis in approximately 2 hr. Both
systems, nonetheless, appear to be useful analytical
tools for tocopherol research.

REFERENCES

1. A. L. Tappel, Ann. N. Y. Acad. Sci. 355, 18 (1980).
2. P. B. McKay and M. M. King, "Biochemical Function--
   Vitamin E: Its Role as Biologic Free-radical Scavenger and Its Relationship to Microsomal Mixed
   Function Oxidase System," in Vitamin E--A Comprehensive Treatise, L. J. Machlin (Ed.), Dekker, New
   York, 1980.
3. J. A. Lucy and A. T. Diplock, FEBS Lett. 29, 205
   (1973).
4. A. S. M. Glasuddin and A. T. Diplock, Arch. Biochem. Biophys. 210, 348 (1981).
5. A. Emmerie and C. Engel, Rec. Trav. Chim. 57, 1351
   (1938).
6. I. D. Desai, "Assay Methods for Tocopherols," in
   Vitamin E--A Comprehensive Treatise, L. J. Machlin
   (Ed.), Dekker, New York, 1980.
7. A. Tsen, Anal. Chem. 33, 849 (1961).
8. M. Kofler, Helv. Chim. Acta 25, 1469 (1942).
9. A. Kouchi, Y. Yaguchi, and G. Katsui, J. Nutr. Sci.
   Vitaminol. 21, 183 (1975).
10. J. F. Cavins and G. E. Inglett, Cereal Chem. 53,
    605 (1974).
11. P. J. Van Niekerk, Anal. Biochem. 52, 533 (1973).

12. R. C. Williams, J. A. Schmitt, and R. A. Henry,
    J. Chromatogr. Sci. 10, 494 (1972).
13. W. O. Landen, J. Chromatogr. 211, 155 (1981).
14. L. J. Hatam and H. J. Kayden, J. Lipid Res. 20,
    639 (1979).

CHAPTER 23

Simultaneous Purification and Analysis of Skin Surface Phospholipids

Tom R. Watkins
Mary E. DiPaola

INTRODUCTION

Analysis of natural products and manufactured products, such as foods and pharmaceuticals, typically entails purification prior to chromatographic separation and detection, whether by TLC, LC, or some other chromatographic method. For example, analysis of polar lipids in a complex mixture such as the phospholipids of normal and psoriatic epidermis has been accomplished after column chromatography to purify the sample lipids and to separate the glycerides and other neutral lipids (1). An enzymatic method may be used to degrade interfering substances. For removal of excess triglyceride, a nutrient emulsion that provides a dietary caloric supplement may be treated with lipase before subsequent vitamin analysis.

In this study, skin surface phospholipids were found to contain several times as much neutral lipid as phospholipid, which prevented routine analysis of the phospholipids. If a skin surface lipid extract was applied to a standard 250 μm layer silica gel TLC plate for analysis without prior purification, waxes and glycerides present at levels severalfold greater than the phospholipids would typically streak across the chromatogram, thereby impairing phospholipid analysis. These compounds at natural concentrations

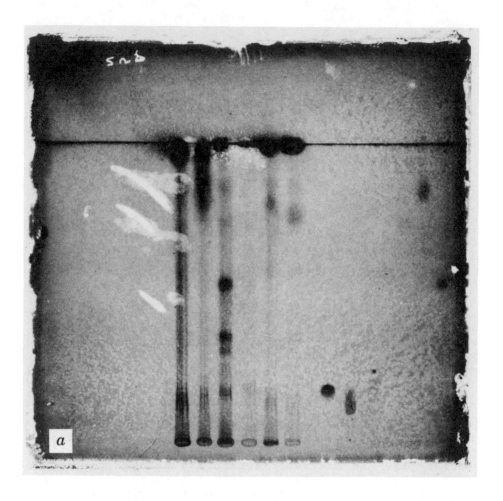

Figure 1 (A). Illustration of the effect of sample
overload.

relative to the phospholipids overloaded the sorbent.
Besides the overloading problem, baseline drift
occurred near the solvent front.   This rising baseline
tended to obscure sample peaks with R_f values near the
front.   The chromatogram in Figure 1(A) illustrates
this problem.   Note the overloaded waxes and glycerides streaked across the phospholipids.   Figure 1(B)

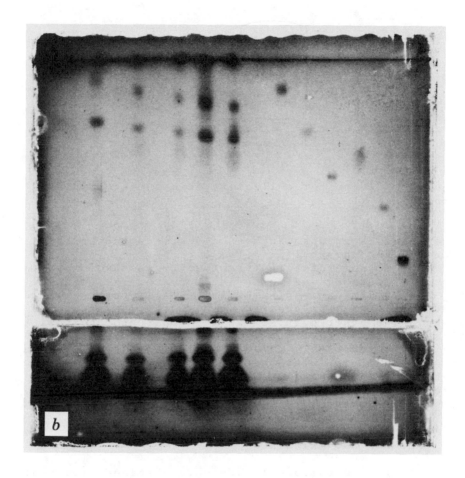

Figure 1 (B). Separation of glycerides from phosphate.

shows glycerides in the lower portion and the phospholipids in the upper portion.
    Other approaches to studies of skin lipids or
skin surface lipids have entailed multiple-solvent
mobile phases or two-dimensional chromatography (2),
paper chromatography requiring several hours for development (3), column chromatography (1, 4), or radio-
label techniques in combination with one or more of
the above. Earlier studies usually separated

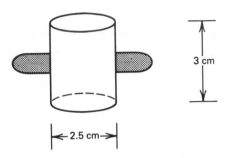

Figure 2. Device for collecting skin surface lipid
samples.

heterogeneous groups, such as glycerides and waxes,
sterols, and phospholipids (5, 6).
     The purpose of the present study was twofold.
The chief purpose was to devise a system for separating the glycerides and waxes from a skin surface lipid
mixture so that the phospholipids were separated as a
group. The other purpose was to compare the composition of skin surface phospholipids of persons consuming a diet relatively limited in lipid, and in particular phospholipid, with that of persons consuming
higher levels. The former was exemplified by omni-
vores, those including meat in the diet, the latter by
herbivores, those forsaking meat.

METHODS AND MATERIALS

Samples of skin surface lipids were obtained from meat-
eating and nonmeat-eating subjects, healthy male and
female college students. Collection of skin surface
lipids was by extraction with heptane-ethanol (3:2 v/v)
with the extractor illustrated in Figure 2. The collection site, either the forehead or the ventral side
of the forearm, was washed with 1 percent Triton X-100,
followed by a distilled water rinse. After waiting
for 1 hr, the lipids excreted on the skin were
extracted for 1 min into 3 ml of solvent. The extract
was dried under nitrogen at 40° C, redissolved in 0.05
ml of chloroform-methanol (19:1 v/v), and frozen until
analysis. Samples were analyzed within two days of
collection.

All solvents used were distilled-in-glass, used
as purchased from Burdick and Jackson (Muskegon,
Michigan). Thin layer plates coated with silica gel G
containing a fluorescent indicator with a 0.25-mm
layer, 20 x 20 cm (Analtech, Newark, Delaware), were
washed by development to the top of the plate with
chloroform-methanol (1:1 v/v) to remove adsorbed
impurities. Activation of the plate after air drying
was done in an oven (100° C) to remove traces of solvent before sample application.

Reference compounds were: lysolecithin (LL),
phosphatidylinositol (PI), sphingomyelin (S), phosphatidylserine (PS), lecithin (L), phosphatidylglycerol
(PG), and phosphatidylethanolamine (PEA). These were
obtained from Supelco (State College, Pennsylvania) or
Biorganics (Newark, Delaware). Reference compounds
and samples were applied to the plate in 0.7 cm bands,
50-4000 ng/band. Calibration curves were constructed
in similar fashion, with four replications for each
mass. After development, plates were air dried
briefly to remove traces of solvent, sprayed to saturation with 10 percent cupric sulfate in 8 percent phosphoric acid, and charred at 190° C for 8 min. Chromatograms were scanned with a densitometer using a deuterium light source (350 nm) (Shimadzu CS-920, Columbia, Maryland).

RESULTS

The two chromatograms obtained from a typical sample
analysis after unidimensional development appear in
Figures 3 and 4. Note particularly the streaked neutral lipid across the chromatogram in Figure 3. This
has been virtually eliminated in Figure 4, as demonstrated by the relatively flat baseline and improved
signal/noise ratio. Figure 5 presents a typical scan
of the neutral lipid removed during the development
in the first direction, which caused the rising baseline in Figure 3 near the front. A scan of a typical
standard curve of pure reference compounds appears in
Figure 6.

Data from the forehead surface phospholipid analyses of persons consuming self-selected, mixed diets,
and others whose diets did not include meat are dis-
played in Table I. The chief phospholipid classes

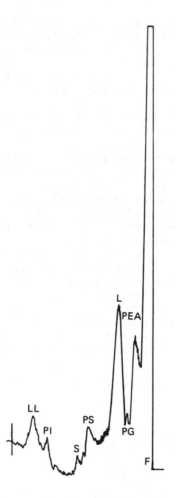

Figure 3. Separation of phospholipids of skin surface
lipids. Note streaking caused by neutral lipid which
results in higher baseline near origin.

measured were, in descending order: lecithin, phosphatidylethanolamine, sphingomyelin, and phosphatidylinositol. Omnivore forehead surface phospholipids contained about twice as much sphingomyelin as herbivores,
18.3 vs. 8.3 percent, but less phosphatidylethanolamine, 16.9 vs. 21.7 percent. Minor phospholipids, on

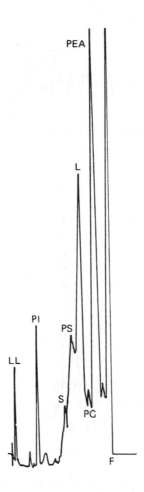

Figure 4. Separation of neutral lipid by development
to region below the starting line with ethyl ether.
The neutral lipid was thus washed away from the sample
and removed by cutting the section off the plate
before development of the phospholipid in the opposite
direction.

a mass basis, were lysolecithin, phosphatidylserine,
and phosphatidylglycerol, present at 4.0, 6.4, and 6.7
percent, respectively, for meat eaters, and 1.2, 9.6,
and 11.6 percent, respectively, for nonmeat eaters.

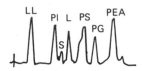

Figure 5. Scan of the neutral lipid zone.

1 / 10 x

Figure 6. Scan of reference phospholipid separated by the described method.

Data measured from lipid samples obtained from the forearm surface phospholipids are presented in Table II. The relative proportion of chief phospholipid classes was not markedly different from the values for the forehead samples. Lecithin and phosphatidylinositol predominated, with somewhat reduced amounts of phosphatidylinositol, phosphatidylserine, and sphingomyelin. Mean values were: 25.9, 18.1, 16.8, 13,1, and 14.0, respectively, for meat eaters;

TABLE I. FOREHEAD SURFACE PHOSPHOLIPID COMPOSITION

				Mass Percent			
	LL	PI	S	PS	L	PG	PEA
Meat Eaters							
x	4.0	17.7	18.3	6.4	26.4	6.7	16.9
SD	4.0	7.9	3.4	4.7	6.0	3.7	3.0
Nonmeat Eaters							
x	1.2	18.9	8.3	9.6	22.8	11.6	21.7
SD	0.7	9.4	5.3	2.3	8.0	3.0	3.4

TABLE II. FOREARM SURFACE PHOSPHOLIPID COMPOSITION

				Mass Percent			
	LL	PI	S	PS	L	PG	PEA
Meat Eaters							
x	0.9	16.8	13.1	14.0	25.9	5.9	18.1
SD	0.5	7.9	7.2	9.9	12.5	2.9	14.0
Nonmeat Eaters							
x	3.1	21.7	11.9	17.5	20.1	9.3	15.8
	3.1	21.7	11.9	17.5	20.1	9.3	15.8

and 20.1, 15.8, 21.7, 11.9, and 17.5, respectively,
for vegetarians. Levels of phosphatidylglycerol were
5.9 for meat eaters and 9.3 for nonmeat eaters.

DISCUSSION

The results presented have proved to be satisfactory
for achieving the first aim, viz., the separation of
neutral lipids, including glycerides and other esters,
from skin surface lipids. This has improved the TLC
analysis of phospholipids in one dimension. The key
to the improved analysis was the removal of less polar
lipids by development of the plate in the opposite
direction of normal development with a low-polarity
mobile phase, breaking the plate below the original
origin, air drying, and development of the phospholipid chromatogram in the usual direction. As the
sample chromatograms showed, the removal of a majority
of neutral lipids facilitated analysis of phospholipid
with less interference in the form of background chromatogram noise and severe baseline rise near the solvent front.
     In contrast to earlier methods with a 10-hr or
longer development time on paper, for example, the TLC
method may be done in less than 1 hr. Since the analyses are done in one dimension, several samples may be
chromatographed on a single plate in the same analysis.
     One further advantage of this method is that the
scored line on the TLC plate may be located in such a
position that both the nonpolar lipids and polar phospholipids may be analyzed on either half of the same
plate. In the present report, the phospholipids were
of greater interest and received greater attention.
     The phospholipid class values measured were not
significantly different for persons including or
excluding meat from a freely selected diet. However,
some differences were noted. A limited number of sample donors were sampled; hence, one must be wary in
making inferences about a larger population from these
data.
     The values reported here for meat-eating subjects
can be compared with results reported for skin phospholipid composition. Difficulties arise because of tissue sampled, skin vs. the surface of the skin, and the
methodological approach, total phospholipid, or the

polar lipid being combined with the neutral lipid.
Ferrando et al. (7) have reported values for normal
skin glycerides, phospholipids, and cerebrosides
together. The relative proportions of the phospholipid values we report are similar with our lecithin
values lower and phosphatidylinositol higher, both for
forehead and forearm lipids. Phosphatidylcholine and
phosphatidylethanolamine values higher than ours were
reported for the epidermis of normal and psoriatic subjects by Tsamboas et al. (1). Rajka (8) has analyzed
and reported skin surface lipid data of subjects with
atopic dermatitis, but the data were for nonhomogeneous classes, such as glycerides, sterols, wax esters,
and phospholipids. Similar class lipid data have also
appeared from the work of Bouissou et al. (6). In
general the data of our meat eaters did not differ
largely from the reported values of normal subjects,
except as noted.

The results of analyses of skin surface lipids
from vegetarian and nonvegetarian subjects allow comparison of dietary effects of feeding a diet relatively limited or rich in phospholipid. Wide ranges
of individual values were expected in either group,
partly since biochemical individuality would also be
manifested in skin phospholipid composition. In addition, some of those who had omitted meat from their
diets had been vegetarians for a mere 2 yr in a few
cases, thus reducing the magnitude of diet upon skin
surface lipid profile. Further, some subjects, though
vegetarian, may have consumed a larger proportion of
processed foods that contain less phospholipid than
minimally processed foods. This may have reduced the
magnitude of the differences in the data of the two
groups according to Watkins (9). As soon as data have
been reported with larger numbers of subjects in each
group, a better estimate of the variance will make possible firmer conclusions about these differences in
skin surface phospholipids.

Gray and Yardley (10) have compared the epidermal
lipids of the human, pig, and rat. The general proportions of phospholipid classes showed similar trends
with some species idiosyncrasies. Their data for lecithin, as well as the data of Tsamboas et al. (1), were
somewhat higher than our data, with phosphatidylinositol and phosphatidylserine accordingly lower. Such
differences may largely reflect the inherent

differences between lipid class turnover rates in normal skin and subsequent excretion. Our data represented the excreted lipids chiefly.

Several diseases of the skin and body lipid
metabolism result in modified skin surface lipids.
This suggests that the skin lipid turnover occurs
sufficiently rapidly to allow sampling as in our
method to be used as a potential indicator of body
tissue status with regard to lipid intake. Such a
noninvasive method of obtaining tissue samples and
the TLC method presented should facilitate further
studies of dietary and metabolic effects upon skin
lipid turnover.

SUMMARY

The purpose of this report was to purify and quanti-
tate skin surface polar lipids without moving neutral
lipids to the front, thereby marring the chromatogram
with interfering, extraneous lipid. By use of a plate
prescored perpendicular to the development direction,
developing the plate in one direction, breaking the
plate, drying, and developing normally, satisfactory
analysis of phospholipid was accomplished. This represented a rigorous chromatographic challenge: separating a minor constituent in a complex mixture. This
has been illustrated by the analysis of skin surface
phospholipids naturally present in the matrix of glycerides and wax esters.

REFERENCES

1. D. Tsamboas, A. Kalofoutis, J. Stratigos, C. Miras,
   and J. Capetanakis, Brit. J. Dermatol. 97, 135
   (1977).
2. B. von Rüstow, D. Metz, D. Kunze, and H. Meffert,
   Dermatol. Monatsschr. 166, 96 (1980).
3. W. Gerstein, J. Invest. Dermatol. 40, 105 (1963).
4. C. Jelenko III, M. L. Wheeler, and T. H. Scott,
   J. Trauma 12, 968 (1972).
5. P. Gross and B. M. Kesten, N. Y. State J. Med.
   2683 (1950).
6. H. Boissou, M. T. Pieraggi, M. Julian, L. Douste-
   Blazy, and J. C. Thiers, Arterial Wall III, 127
   (1976).

7. J. Ferrando, V. and V. Corominas, Med. Cut. I.L.A.
    1, 63 (1976).

8. G. Rajka, Arch. Derm. Forsch. 251, 43 (1974).

9. T. Watkins, unpublished data, 1983.

10. G. M. Gray and H. J. Yardley, J. Lipid Res. 16,
    434 (1975).

11. G. Nelson, "Fractionation of Phospholipids," in
    Analysis of Lipids and Lipoproteins, E. Perkins
    (Ed.), American Oil Chemists Society, Evanston,
    Ill., 1975.

CHAPTER 24

Think Layer Chromatography of Nucleotides and Nucleosides

Barry P. Sleckman
Joseph C. Touchstone
Joseph Sherma

INTRODUCTION

This paper describes the results of a comparative
study of various TLC systems for the separation of
four representative nucleotides and their corresponding nucleosides. The systems included conventional
and high-performance silica gel; C_2 and C_{18} chemically
bonded reversed phase; and chemically bonded strong
cation exchanger layers, with mobile phases that were
completely aqueous, aqueous-organic, or completely
organic and had various pH and ionic strength values.
The mobile phases used were chosen after preliminary
testing of numerous promising systems described in the
literature (1-4). Retention data, separations, and
possible mechanisms on the different layers are presented.

MATERIALS AND METHODS

The following nucleotide and nucleoside standards were
obtained from commercial sources:
    Nucleotides
          A. Adenosine-5-monophosphate
          B. Uridine-5-monophosphate

C. Cytidine-5-monophosphate
D. Guanosine-5-monophosphate

Nucleosides

E. Adenosine
F. Uridine
G. Cytidine
H. Guanosine

Standards were prepared in aqueous solutions at concentrations of 3.0 and 0.3 μg/μl. Solutions were applied to layers with a Drummond Digital Microdispenser at levels ranging from 1 to 30 μg.

The TLC systems used were as follows:

System 1. Whatman LK5DF, 20 x 20 cm, preadsorbent silica gel plates; isopropanol-water-ammonia (7:2:1 v/v) mobile phase. Development time 3.5 hr. Standards applied at level of 15 μg.

System 2. Whatman LHPKDF, 10 x 10 cm, preadsorbent high-performance silica gel plates; isopropanol-water-ammonia (7:2:1 v/v) mobile phase. Development time 1.75 hr. Standards applied at 1 μg.

System 3. Whatman $LKC_{18}DF$, 20 x 20 cm, preadsorbent octadecylsilane reversed-phase plates; isopropanol-water-DMSO (16:4:1 v/v) mobile phase. Development time 4 hr. Standards applied at 30 μg.

System 4. Whatman C_2, 20 x 5 cm, dimethylsilane reversed-phase plates; isopropanol-water-DMSO (16:4:1 v/v) mobile phase. Development time 3.5 hr. Standards applied at 18 μg.

System 5. Whatman SCX, 20 x 20 cm, polymer bound, chemically bonded, SO_3^- strong cation exchange plates, NH_4^+ form; 0.1 M $NH_4H_2PO_4$-1.0 M NaCl mobile phase. Development time 1 hr. Standards applied at 18 μg.

System 6. Whatman SCX, 20 x 20 cm, strong cation-exchange plates, NH_4^+ form; 1.0 M NaCl mobile phase. Development time 1 hr. Standards applied at 18 μg.

System 7. Whatman SCX, 20 x 20 cm, strong cation exchange plates; NH_4^+ form; isopropanol-water-ammonia (7:2:1 v/v) mobile phase. Development time 4.5 hr. Standards applied at 18 μg.

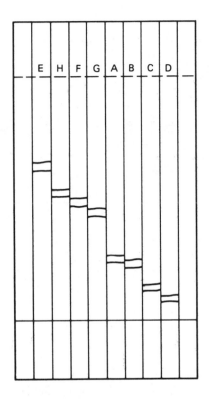

Figure 1. Separation of nucleotides and nucleosides (compounds A-G) on a silica gel layer (TLC System 1).

System 8. Same as System 1, but development time 4.5 hr, standards applied at 18 μg.

Chromatograms in TLC Systems 1-4 were developed for a distance of 13.5 cm past the origin (sorbent-preadsorbent interface), except TLC System 2, which was developed 7 cm past the interface. Chromatograms in TLC Systems 1-4 were developed in rectangular tanks that were paper lined and allowed to equilibrate with the mobile phase for 10 min prior to inserting the plate. Chromatograms in TLC Systems 5-8 were developed for a distance of 16 cm past the origin in unlined, unequilibrated rectangular tanks. Development times are given above.

TABLE I. R_f RANGES OF COMPOUNDS

TLC Systems

Compound	1	2	3	4	5	6	7	8
A	23.6-26.4	35.3-43.1	46.4-52.1	55.0-63.6	66.3-72.5	58.1-65.0	13.1-17.5	27.5-31.3
B	21.4-25.0	35.4-42.8	55.0-58.6	72.1-74.3	89.4-93.8	86.3-91.9	18.8-23.1	25.6-30.0
C	12.1-14.3	27.4-31.2	27.9-32.1	21.4-31.4	82.5-86.3	75.6-80.0	29.4-35.0	16.9-21.3
D	7.9-10.0	21.5-27.7	20.0-24.3	17.9-22.1	79.4-84.4	73.8-78.8	3.8-8.1	11.9-16.3
E	60.7-65.0	73.8-80.0	64.3-67.9	77.9-82.9	65.0-70.0	58.8-63.1	56.3-63.1	58.8-64.4
F	46.4-50.0	61.5-67.7	63.6-67.9	80.7-85.0	92.5-97.5	85.0-90.6	43.8-48.1	48.1-52.5
G	42.1-45.7	52.3-58.4	57.1-60.0	72.9-76.4	79.4-83.8	75.6-79.4	32.5-36.3	41.9-46.3
H	50.7-53.5	67.9-71.9	57.9-60.7	72.1-75.7	79.4-83.8	73.1-77.5	36.9-42.5	51.3-55.0

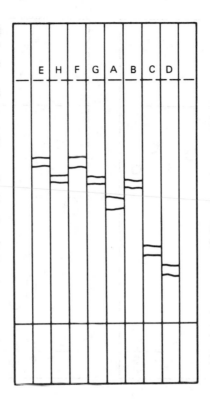

Figure 2. Separation of nucleotides and nucleosides
(compounds A-G) on a C_{18} reversed-phase layer (TLC
System 3).

        Visualization was by fluorescence quenching under
a shortwave UV lamp. All layers contained fluorescent
indicator and were activated at 254 nm.

        RESULTS AND DISCUSSION

Table I presents the R_f values of the rear and front
limits of the compound zones in the eight TLC systems
studied. Silica gel System 1 allowed the separation
of six of the eight compounds (Figure 1). The use of

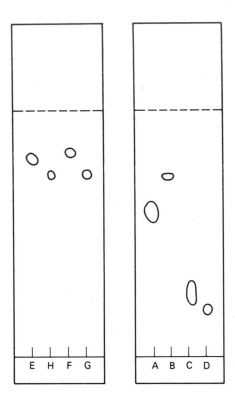

Figure 3. Separation of nucleotides and nucleosides (compounds A-G) on C_2 reversed-phase layers (TLC System 4).

high-performance silica gel thin layers (TLC System 2) increased resolution but did not increase the number of compounds that could be separated (Table I) since compounds A and B were not fully resolved on either layer. Development was faster on high-performance layers and minimum detectable levels of the nucleotides and nucleosides were decreased compared to the regular silica gel layers. The minimum detection level on the high-performance layers was approximately 1 µg, whereas the minimum detection level on the regular silica gel layers was approximately 15 µg. The level of detection could be decreased considerably if

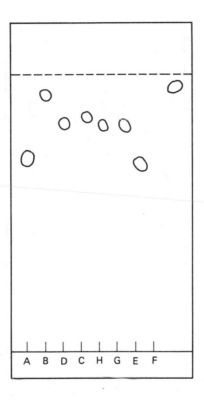

Figure 4. Separation of nucleotides and nucleosides
(compounds A-G) on a SCX ion-exchange layer developed
with buffer (TLC System 5).

the N-chlorination detection procedure described by
Schwartz was used (5). The nucleotide and nucleoside
standards used in this study were detected at levels
as low as 100 ng in spot tests using this reagent.
However, the reagent was found to be unsuitable for
detecting nucleotides and nucleosides that were developed in TLC systems employing amine solvents in the
mobile phase. Since the reversed phase thin layer systems were the only ones that did not require an amine
component in the mobile phase, only these would allow
use of the more sensitive detection reagent.
     TLC system 3 gave separation of only five of the
eight compounds studied (Table I and Figure 2).

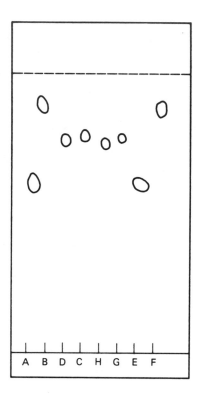

Figure 5. Separation of nucleotides and nucleosides
(compounds A-G add on an SCX ion-exchange layer developed with salt (TLC System 6).

However, this system was capable of resolving compounds C and D, G and F, and A and B, which were difficult to resolve in TLC System 1. The DMSO in the
mobile phase of System 3 was needed to tighten the
bands, causing a significant increase in resolution
compared to results without DMSO.
      TLC System 1 in combination with TLC System 3
gave separation of all compounds studied. That is to
say, if two compounds are unresolvable in System 1,
they are resolved in System 3, and those that are unresolvable in System 3 are resolvable in System 1. All
the nucleotide standards, in both TLC Systems 1 and 3,
can be resolved from their respective nucleosides.

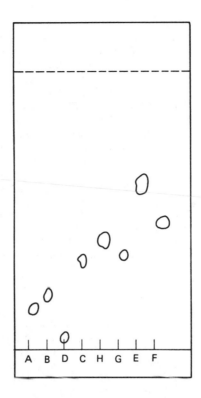

Figure 6. Separation of nucleotides and nucleosides (compounds A-G) on an SCX ion-exchange layer developed with mixed aqueous-organic solvent (TLC System 7).

The results of Systems 3 and 4 (Table I and Figure 3) show that the mechanism of separation of nucleotides and nucleosides on reversed-phase thin layers is not true reversed phase. Although the order of some compounds is reversed, the overall order of separation is by no means the complete opposite of that observed on silica gel thin layers. Some of the compounds that occupy relatively low R_f ranges on silica gel thin layers showed higher R_f in the reversed-phase system, but those that occupy relatively high R_f ranges remain at these positions. The absence of a reversed-phase mechanism for amino acid separations on C_{18} layers has been demonstrated earlier (6).

Nucleotides and nucleosides were also developed on silica-based strong acid cation-exchange layers (System 5, Table I and Figure 4). The pH was varied from approximately 3.0 to 9.0 with HCl or NH_4OH in order to decrease the degree of migration. However, altering the pH in this manner did not result in changes in spot migration, as the compounds constantly occupied high R_f ranges. Systems with varying concentrations of $NH_4H_2PO_4$ and NaCl were also tested. The concentration of $NH_4H_2PO_4$ was varied over a range of 0.0 to 0.5 M with no appreciable change in spot migration. At the same time, the NaCl concentration was varied over a range of 0.0 to 2.0 M, also with no appreciable change in spot migration. When the NaCl concentration was below 0.75 M, smearing and streaking of all nucleotides and nucleosides occurred at the solvent front, such that none of them could be identified. It is believed that this is due to the fact that below a certain ionic strength, the buffer may wash the binder out of the layer.

When the nucleotide and nucleoside standards were developed in System 6 (Table I and Figure 5) with a 1.0 M NaCl mobile phase, the standards migrated in the same manner as they did in the buffer systems. These observations, namely the lack of change in spot migration with varying buffer pH and net ionic concentration, suggest the absence of a true ion-exchange mechanism.

In order to determine if the silica gel in the SCX layer was affecting the separation of the compounds studied, further runs with mobile phases used for the separation of nucleotides and nucleosides on silica gel were carried out (System 7, Table I and Figure 6; and System 8, Table I and Figure 1). The data and figures show that the compounds had relatively the same order of separation on both the silica gel and SCX thin layers when developed in the same mobile phase. The development times for the respective systems were also the same. These two observations further substantiate the thought that the silica gel is mainly involved in the mechanism of separation of the compounds studied on the SCX layers.

REFERENCES

1. R. Cardinaud, J. Chromatogr. 154, 345 (1978).
2. R. Cardinaud and J. Holguin, J. Chromatogr. 115, 673 (1975).
3. H. J. Issaq, E. J. Barr, and W. L. Zielinski, J. Chromatogr. 131, 265 (1977).
4. J. Tomasz, Chromatographia 13, 469 (1980).
5. D. P. Schwartz and J. T. Sherman, J. Chromatogr. 240, 206 (1982).
6. J. Sherma, B. P. Sleckman, and D. W. Armstrong, J. Liq. Chromatogr. 6, 95 (1983).

CHAPTER 25

Direct Microanalysis of Lipids in Cultured Cells by Thin Layer Chromatography

Howard R. Sloan
Constance S. Seckel

INTRODUCTION

Studies of lipid metabolism in cultured cells are complicated by the time and expense required to obtain more than a few milligrams of tissue for investigation. Frequently, one wants to analyze rates of lipid synthesis in as few as 10 to 100,000 cultured cells, a number that corresponds to 10 to 100 μg of tissue protein. To do so with currently available techniques is, at best, very difficult. The limited sample size makes it difficult to perform such standard lipid extraction techniques as those described by Folch or Dole (1, 2). Although these techniques can be scaled down so that milligram quantities of tissue can be studied, even careful attention to details will not prevent the incurring of relatively large errors. The major sources of these errors are the multiple transfer and concentration steps which are associated with losses of material on vessel surfaces. In addition to problems with quantitation when standard techniques are employed at the microlevel, such methods are also extremely time-consuming and tedious. For example, the first step in the standard techniques for lipid analysis is microextraction, a procedure that takes as long as 8 to 12 hr. Thereafter, the nonlipid material must be removed from the extract by a partition step which takes an additional 2 to 12 hr. Finally, the sample must be concentrated prior to fractionation or analysis by either column or thin layer chromatography.

This paper describes a thin layer technique that permits the simultaneous extraction and lipid analysis of a large number of cell lines. By suitably modifying the technique, it is also possible to quantify the rate of lipid synthesis in these cell lines. The method requires as few as 1000 cultured cells and takes less than 15 min of manipulation per cell line. The technique is made possible by the availability of thin layer chromatographic plates with a special preadsorbent area composed of diatomaceous earth which permit the direct application of lysates of cultured cells without resorting to prior solvent extraction, partition, transfer, and concentration of the samples. Development of the chromatographic plate with the appropriate solvents simultaneously extracts the lipids from the cellular material, precipitates the cellular proteins, and separates the various lipid classes. In this report, we describe our application of this technique to the analysis of both the neutral and polar lipids in small samples of cultured human cells.

MATERIAL AND METHODS

A model 2955 Transidyne Scanning Densitometer (Transidyne, Ann Arbor, Michigan) equipped with a Spectro-Physics Minigrator (Santa Clara, California) was employed for scanning the thin layer plates. The plates and radioautograms were scanned at 600 nm in the reflectance mode with a 9 mm-long and 0.8 mm-wide beam. The Minigrator calculated peak areas. For thin layer chromatography, Whatman (Clifton, New Jersey) LK5D plates were used; these plates have a 250 µm-thick layer and are prescored into 9 mm lanes.

All lipid standards were purchased from either Applied Science Laboratories (State College, Pennsylvania) or Supelco, Inc. (Bellefonte, Pennsylvania). They were assayed for purity by thin layer chromatography before use and were at least 99 percent pure. Sodium dodecyl sulfate was purchased from Bio-Rad Laboratories, (Richmond, California), Cutscum® from Fisher Scientific (Pittsburgh, Pennsylvania), and

Triton® X-100 from Packard Instruments, Inc. (Downers Grove, Illinois). The x-ray film, Kodak® OG-6, Ortho G, was the product of Eastman Kodak Corporation, Rochester, New York). Scinti Verse Bio HP scintillation fluid was purchased from Fisher Scientific (Pittsburgh, Pennsylvania). All radioactive chemicals were purchased from New England Nuclear (Boston, Massachusetts). All other chemicals were reagent grade; the solvents were not redistilled prior to use.

Fibroblast-like tissue culture cell lines were initiated from 3 mm pieces of skin obtained from normal volunteers or from samples of human foreskin. The cultures were initiated and propagated as described previously (3). The tissue culture media employed were Ham's F-10 with glutamine and Ham's F-10 without choline or methionine; the methionine and choline concentrations of the latter medium were adjusted to that present in standard Ham's F-10 by the addition of ^{14}C-methyl choline (specific radioactivity, 1 µCi/ µmol). In experiments in which the uptake of phosphate into phospholipids was assessed, sufficient ^{32}P-sodium phosphate was added to the culture medium to adjust the specific radioactivity to 3 µCi/µmol. In experiments in which the uptake of sodium acetate into all lipids was assessed, sufficient C-14 sodium acetate was added to the culture medium to adjust the specific radioactivity to 3 µCi/µmol. The fibroblasts were grown in 75 cm^2 plastic vessels. They were harvested from the plastic surface by the addition of a small volume of 1 percent trypsin followed by brief incubation at 37° C (3). The harvested cells were washed three times with Dulbecco's phosphate buffered saline without calcium or magnesium; following each wash, the cells were sedimented by centrifugation at 600 x g. The cell pellet was suspended in a minimal volume of distilled water, approximately 20 µl/10^6 cells, and the cells were lysed by the addition of four volumes of either Triton® X-100 (25 gm/dl), Cutscum® (25 gm/dl), or sodium dodecyl sulfate (5 gm/dl) so that the final concentrations of detergents were 5, 5, and 1 gm/dl, respectively. The progress of the lytic process was monitored in representative cultures by observing the cell suspension under an inverted phase contrast microscope. In some experiments, the lysis was accomplished by the addition to the cell pellet suspension of four volumes of distilled water

followed by freezing and thawing the suspension three
times.

The fibroblast lysates are readily maintained in
a relatively uniform state of dispersion by periodic
agitation on a Vortex mixer. Small volumes, 5 to 20
µl of the cell suspension, are applied to the preadsorbent area of the Whatman LK5D plates, 10 mm from the
silica gel-diatomaceous earth interface. The plates
are dried thoroughly with a jet of warm air from a
hair dryer and developed with either a polar (chloro-
form:methanol:water, 65:25:4) or a nonpolar (petroleum
ether:ethyl ether:glacial acetic acid, 90:10:1) solvent (4). After development, the plates are again
dried thoroughly. The various lipids are detected by
exposing the plates to iodine vapor, by staining the
thin layer plates with an anisaldehyde reagent (4), or
by staining them with the cupric acetate stain
described by Touchstone et al. (5).

In those experiments designed to determine the
lipid content of cultured fibroblast-like cells, the
plates are scanned with the Transidyne Scanning Densitometer, following staining with the cupric acetate
stain, as described previously (6). In those experiments in which the neutral lipids (cholesterol, cholesteryl ester, and triglyceride) were quantified, methyl
oleate, which migrates between cholesteryl ester and
triglyceride, can be employed as an internal standard
at a concentration of 5 mg/ml. In most experiments,
plasma samples, with known concentrations of the various lipids, were applied to lanes adjacent to those on
which the unknown samples had been applied. The peak
area corresponding to the color density of each lipid
spot was determined by the Minigrator and compared to
the area obtained from known amounts of the same lipid
in the control lanes (6).

In those experiments in which the rates of synthesis of the various lipid are being investigated, radioautograms are prepared by placing the Ortho G film on
the developed thin layer plates in standard, spring-
loaded x-ray cassettes. The plates are exposed to the
x-ray film for various lengths of time at room temperature. Following development of the x-ray film, the
thin layer plates and the developed x-rays are super-
imposed. The area of the thin layer plate corresponding to the lipid under investigation is identified and
the silica gel from that area of the plate is removed

by scraping the area with a razor blade. The powder
is placed into a 20 ml glass scintillation vial; 10 ml
of the scintillation fluid are added, and the bottles
are briskly agitated with a vortex mixer. The samples are then counted in a Beckman Model LS 7500
Liquid Scintillation Spectrophotometer. From lanes to
which no radioactive material had been applied,
approximately equal amounts of silica gel are scraped
and treated in a similar fashion in order to determine
the background radioactivity of the plates.

RESULTS AND DISCUSSION

In preliminary experiments, we observed that attempt-
ing to lyse cultured cells by osmotically shocking
them in distilled water, and then repeatedly freezing
and thawing them is unreliable. This method produces
lysates which are extremely nonhomogeneous in nature;
moreover, the cellular fragments in such preparations
are of relatively large size. On the other hand, each
of the three detergents employed, that is, Triton®
X-100, Cutscum®, and sodium dodecyl sulfate, provides
almost instantaneous lysis of the fibroblasts. The
resulting lysates contain cell fragments which are of
extremely small size when viewed under phase micros-
copy. Moreover, the resulting suspensions are quite
stable and require only periodic agitation in order to
maintain the preparation in a uniform state of dispersion. Although each of the detergents effectively
lyses fibroblasts, they are organic compounds and
might therefore, at least theoretically, interfere
with the thin-layer chromatographic separation of cellular lipids. Both Cutscum® and Triton® X-100 barely
migrate into the silica gel portion of the thin layer
plates when the nonpolar solvent is employed; both of
these detergents move to the solvent front when the
polar solvent is used. On the other hand, sodium
dodecyl sulfate does not leave the diatomaceous earth
portion of the plate when the nonpolar solvent is
employed, but has an R_f of 0.65 in the polar solvent.
This property of sodium dodecyl sulfate makes it of
limited usefulness if quantitative spectrophotodensitometry is to be employed with polar lipids. We found
that, at the concentrations employed in these studies,
none of the three detergents affects either the R_f of

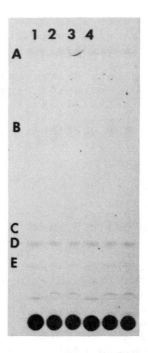

Figure 1. Radioautogram prepared from thin layer chromatographic separations of the nonpolar lipids in
human diploid fibroblasts.   Lanes 1 and 2 contain
lysates prepared with Triton® X-100; lanes 3 and 4 contain lysates prepared with sodium dodecyl sulfate.
The solvent used was petroleum ether:ethyl ether:
glacial acetic acid, 90:10:1.   (A) Cholesteryl esters;
(B) triglyceride; (C) fatty acids; (D) cholesterol.

an individual lipid class nor the separation of one
lipid from another.   In addition, the presence of
detergent in the sample does not produce broadening of
the spots associated with any lipid class.
     A typical radioautogram depicting the separation
of the nonpolar lipids in a fibroblast cell line is
depicted in Figure 1.   The fibroblasts were grown for
three days in the presence of ^{14}C-sodium acetate, as
described above.   An aliquot of the lysate, corresponding to approximately 5-10,000 cells, was applied

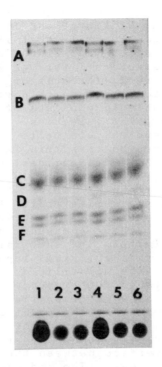

Figure 2. Radioautogram prepared from thin layer chromatographic separations of the polar lipids in human
diploid fibroblasts. Lanes 1 and 2 contain lysates
prepared with Triton® X-100; lanes 3 and 4 contain
lysates prepared with Cutscum®; lanes 5 and 6 contain
lysates prepared with sodium dodecyl sulfate. The
solvent used was chloroform:methanol:water, 65:25:4.
(A) Nonpolar lipids; (B) phosphatidyl serine; (C)
phosphatidyl choline; (D) phosphatidyl inositol; (E)
sphingomyelin; (F) lysophosphatidyl choline.

to the diatomaceous earth portion of the thin layer
plates; the plates were then developed with the nonpolar solvent. The location of the various lipid
classes was determined by comparison with the position
to which lipid standards migrated in adjacent lanes.
The adjacent lanes were usually stained with iodine
vapors. Occasionally, the standard lanes were cut
from the glass plate and stained with the anisaldehyde

reagent. The amount of fibroblast lipid applied to
each of the lanes was so small that it was beyond the
range of detection of the anisaldehyde reagent which
can conveniently detect as little as 100 ng. As
illustrated in Figure 1, the bulk of the radioactivity
remains in the diatomaceous earth portion of the plate.
This material undoubtedly consists of unreacted sodium
acetate and highly polar molecules formed from sodium
acetate by the tissue culture cells. The major nonpolar lipids resolved on this thin layer are: A,
cholesteryl ester; B, triglycerides; C, nonesterified
fatty acids; and D, cholesterol. In addition there
are at least two more polar lipids which migrate
between the origin and cholesterol. In this study,
these lipids were not identified; based on visual
inspection, however, these compounds appear to be synthesized at approximately the same rate as fatty acids.
       Figure 2 is a radioautogram prepared from a thin
layer separation of the polar lipids of fibroblasts
grown for three days in the presence of ^{14}C-sodium
acetate. Once again, a major portion of the radioactive material remains at the point of sample application. An additional small amount of radioactivity
(the sharp band directly below the arabic numerals)
migrates to the interface between the diatomaceous
earth and the silica gel. The radioactive material
migrating with the solvent front probably represents
the nonpolar lipids which were not identified on this
chromatogram. B denotes phosphatidyl serine, C represents phosphatidyl choline, whereas D denotes phosphatidyl inositol. E denotes the typical doublet that is
routinely seen with sphingomyelin, whereas F represents lysophosphatidyl choline. The radioactive lipids are well separated from each other and comigrate
precisely with standard lipids applied to adjacent
lanes (not illustrated in this figure). Lanes 1a and
2 contain the lipids in lysates prepared with Triton®
X-100, lanes 3 and 4 contain the fibroblasts prepared
with Cutscum®, and lanes 5 and 6 contain lysates prepared with sodium dodecyl sulfate. Even sodium
dodecylsulfate, which migrates at approximately the
same rate as phosphatidyl choline, did not significantly alter the separation of the various lipid
classes. The more rapid migration of the lipids in
the higher numbered lanes probably represents an edge
effect.

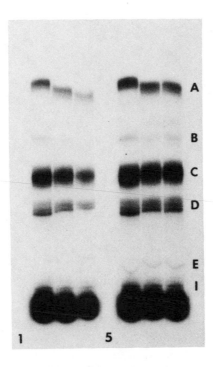

Figure 3. Radioautogram prepared from thin layer chromatographic separations of the polar lipids in human
diploid fibroblasts. Lanes 1 and 5 contain the standards which were not radioactive. Lanes 2, 3, and 4
contain lysates prepared with Triton® X-100 from 1000
cells; lanes 6, 7, and 8 contain lysates prepared from
5000 cell with Triton® X-100. The solvent used was
chloroform:methanol:water, 65:25:4. (A) Nonpolar lipids; (B) phosphatidyl serine; (C) phosphatidyl choline;
(D) phosphatidyl inositol; (E) unidentified; (I) silica gel-diatomaceous earth interface.

        Figure 3 depicts a radioautogram prepared from a
thin layer separation of the polar lipids of fibroblasts grown for 24 hr in the presence of ^{32}P-sodium
phosphate; the radioautogram was deliberately over-
exposed in order to detect the formation of small
amounts of lipids. Lanes 1 and 5 contained lipid
standards against which the ^{32}P-containing lipids were

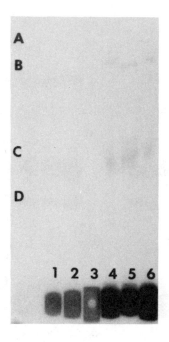

Figure 4. Radioautogram prepared from thin layer chromatographic separations of the polar lipids in human diploid fibroblasts.   Lanes 1, 2, and 3 contain lysates prepared from 5000 cells grown in ^{32}P-sodium phosphate for 1 hr; lanes 4, 5, 6 contain lysates 5000 cells grown in ^{32}P-sodium phosphate for 3 hr.   The solvent used was chloroform:methanol:water, 65:25:4. (A) Triton® X-100; (B) phosphatidyl serine; (C) phosphatidyl choline; (D) sphingomyelin.

compared for identification purposes. An amount of lysate corresponding to 1000 cells was applied in lanes 2, 3, and 4 and to 5000 cells in lanes 6, 7, and 8.   Once again, a large percentage of the radioactivity remains at the origin with lesser amounts being distributed among the various phospholipid classes. The sphingomyelin doublet (band D) is somewhat difficult to identify in this radioautogram, because of the prolonged exposure of the x-ray film to the thin layer plate.   The phosphate-containing material denoted by

the letter E was not identified; in addition, there
was a significant amount of radioactive phosphous-
containing material at the diatomaceous earth-silica
gel interface.

Figure 4 is a radioautogram prepared from a thin
layer separation of the polar lipids of fibroblasts
grown in the presence of ^{32}P-sodium phosphate.   An
amount of lysate corresponding to 5000 cells was
applied to each lane.   The cells had been grown in
^{32}P-sodium phosphate containing medium for 1 hr (lanes
1, 2, and 3) or 3 hr (lanes 4, 5, and 6).   There was
relatively little labeling of phospholipid after only
1 hr of growth, but by 3 hr significant amounts of
phosphatidyl serine (B) and phosphatidyl choline (C)
had formed.   On the original radioautograms, a small
amount of sphingomyelin (D) could also be detected,
although its presence is not visible in these photographs.   The location of the Triton® X-100 which was
used to prepare these lysates is denoted by A.

The extent to which the lipid content of the cultured cells was extracted by the techniques described
in this paper was evaluated by two methods.   First, in
experiments in which lipids were labeled by incubation
of the cells in ^{14}C-methyl choline and ^{14}C-methyl methionine containing medium or in medium containing ^{32}P-
sodium phosphate, the material at the origin was subjected to a second chromatographic development with
the same solvent that had been employed in the first
development, that is, either the polar or the nonpolar
solvent.   At the end of 2 hr of development, the upper
1 cm of the diatomaceous earth portion of the thin
layer plates was scraped from the glass surface and
its radioactive content determined by scintillation
spectrophotometry.   In no case was more than an additional 1 to 2 percent of radioactive material removed
from the diatomaceous earth portion by this second
development with the solvent.   A second line of evidence suggesting that effective extraction and resolution of cellular lipids is achieved was provided by
scraping areas of the plate between the bands of the
major lipid classes.   These areas contained very small
amounts of radioactivity; in no case did the silica
gel in the intervening spaces contain more than 2 percent of the radioactivity in the adjacent major band.
If continuing extraction of the lipids within the cell
were occurring throughout the development of the thin

layer chromatogram, one would expect smears rather
than bands, and therefore one would find a continuous
array of radioactive material along the length of
each lane. Additional evidence that effective, and
essentially complete, extraction and separation of cellular lipids is achieved by this technique is provided
by the excellent resolution of the various lipid
classes demonstrated in Figures 1 through 4. We routinely observe the classical sphingomyelin doublet;
this separation depends on the ability of thin layer
chromatography to separate sphingomyelines containing
hydroxy fatty acids from sphingomyelines containing
all other types of fatty acids. If extraction were
incomplete or were occurring on an ongoing basis
throughout the time that solvent was migrating up the
thin layer plate, one would not expect the production
of sharp bands.

     It seems quite likely that a combination of desir-
able forces promotes the complete extraction and separation of the lipids within cultured cells by the
techniques described in this manuscript. First, lysis
of the cells by osmotic shock ruptures cellular and
subcellular membranes into small pieces which can be
readily extracted by the organic solvent mixtures
employed in thin layer chromatography. Second, the
partial dissolution of the lipid membranes by the
detergents probably renders them more susceptible to
the lipid solvents employed in the separation steps.
Finally, thin layer chromatography provides a large
surface area which should enhance lipid extraction.

     The method for lipid analysis presented in this
manuscript is commended by its simplicity, its rapid-
ity, and by its requirement for relatively small
amounts of sample. It is possible to use the technique to determine both the qualitative content of
lipid classes within cultured cells and the rate of
synthesis of such lipids by cells in culture. The
technique particularly lends itself to the analysis of
the lipid composition and the rates of lipid synthesis
in large numbers of cell lines or in very small
amounts of cultured cells.

ACKNOWLEDGMENTS

This research was supported by Grants 74-351 from the
Central Ohio Heart Chapter, Inc., 74-228 from the
Columbus Children's Hospital Research Foundation, and
1 RO1 AM29996-01A2 from the National Institutes of
Health.
    We wish to thank Mrs. Rita Compston for her
assistance in preparing the manuscript.

REFERENCES

1. J. Folch, M. Lees, and G. H. Sloane-Stanley, J.
   Biol. Chem. 226, 497 (1957).
2. V. P. Dole, J. Clin. Invest. 35, 150 (1956).
3. H. R. Sloan, B. W. Uhlendorf, C. B. Jacobson, and
   D. S. Fredrickson, Ped. Res. 3, 532 (1969).
4. P. O. Kwiterovich, H. R. Sloan, and D. S. Fredrickson, J. Lipid Res. 11, 322 (1970).
5. J. C. Touchstone, M. F. Dobbins, C. Z. Hersch, A. R.
   Baldino, and D. Kritchevsky, Clin. Chem. 24, 1496
   (1978).
6. H. R. Sloan and C. S. Seckel, in Advances in Thin
   Layer Chromatography, J. C. Touchstone (Ed.), New
   York, Wiley, chapter 15, p. 209-218 (1982).

CHAPTER 26

Industrial Applications of Quantitative Thin Layer Chromatography

Laszlo R. Treiber

INTRODUCTION

The general features and the technical specifications
of an instrument are usually well documented by its
manufacturer. It is, however, up to the analytical
chemist to establish the range and limitations of
applicability of every individual method using the
instrument. Such an investigation has been reported
in a recent publication (1) describing an assay method
for Cephamycin C in fermentation broths. The statistical evaluation, including a comparative study with the
routinely used HPLC assay, is discussed there in
details.
       Similar evaluation of every method is required in
our laboratories before we rely on it for making technical decisions. Some selected examples are presented
below, to demonstrate how quantitative TLC (QTLC) has
helped us in the area of industrial process development.

DISCUSSION

Figure 1 demonstrates the usefulness of QTLC in determining the production of Cephamycin C as a function of

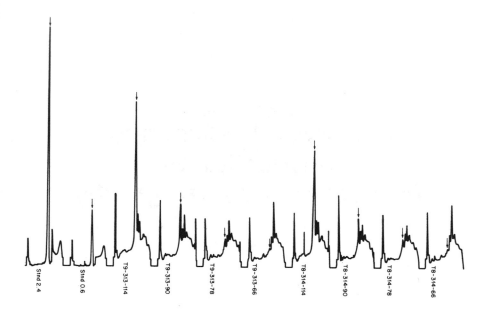

Figure 1. QTLC scan of cephamycin C samples at 273 nm
following chromatography in ethanol-glacial acetic
acid-conc. ammonium hydroxide (6:3:1 v/v) on silica
gel 60 F-254 TLC plates (E. Merck). Instrument:
Shimadzu Model CS-920.

fermentation time.   The arrows mark the antibiotic in
two standards (2.4 and 0.6 g/l) and in two fermentation broths (batch # T9-313 and T8-314) at different
times of the fermentation (the remaining portion of
the numbers indicate fermentation times in hours).
The rise of the peak is directly proportional to the
antibiotic concentration at any given time. Based on
the QTLC assays technical decisions could be made,
such as determining the optimum harvest time of the
batch.
    The very polar developing solvent (1) has caused
layer constituents also to migrate, leading to high
background. However, the conclusions of the assays
were not interfered with at all.
    Another example is a new family of compounds with
anthelmintic activity, called the Avermectins,

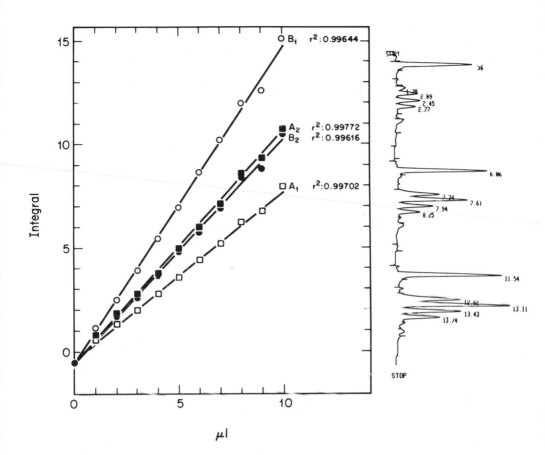

Figure 2. Calibration of avermectins at 245 nm following chromatography in ethyl acetate on silica gel 60
F-254 TLC plates (E. Merck). Instrument: Shimadzu
Model CS-920 combined with a Hewlett-Packard integrator Model 3390A. On the right: automatic QTLC
scan of three broth samples. Origin: 0.56, 6.06,
11.54 min; B_2, 1.79, 7.34, 12.82 min; B_1, 2.09, 7.61,
13.11 min; A_2, 2.45, 7.94, 13.43 min; A_1, 2.77, 8.25,
13.74 min.

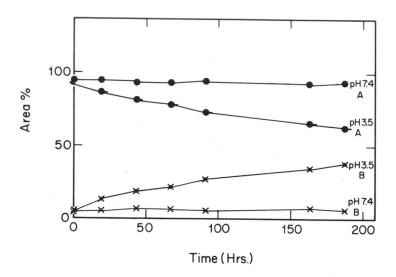

Figure 3. Kinetics of isomerization of the efrotomycin A component to the B isomer at room temperature and pH 7.4 and 3.5.

produced by fermentation of Streptomyces avermitilis MA-4680 (2). The biologically active components, four pairs of compounds as the principal constituents, are structurally related. Their separation has been accomplished by HPLC (3).

The successful resolution of the four pairs by TLC on silica gel 60F-254 (E. Merck) in ethyl acetate as mobile phase opened up the possibility for a quantitative TLC method. Calibration was tested by applying increasing volumes of a whole broth-acetone mixture onto a TLC plate for subsequent development. The calibration was linear for all four components within the entire practical range (Figure 2). The method was found suitable for studying fermentation kinetics, monitoring isolation processes, and with a sample throughput of more than 100 assays per working day in the current set-up, for a variety of screening programs. The decision-making power of the QTLC method, within the range of its applicability, is the same as that of HPLC.

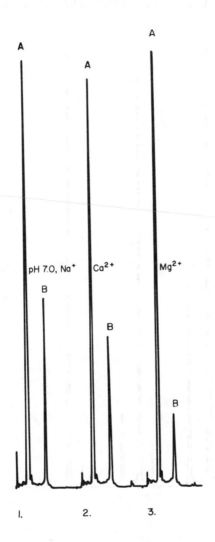

Figure 4. Conversion of the efrotomycin B component to
the A isomer at pH 7.0 in the presence of metal ions.
The samples were evaluated automatically by means of a
Shimadzu Model CS-920 QTLC scanner at 235 nm following
chromatography in chloroform-methanol-conc. ammonium
hydroxide (80:20:1 v/v) on silica gel 60 F-254 TLC
plates (E. Merck).

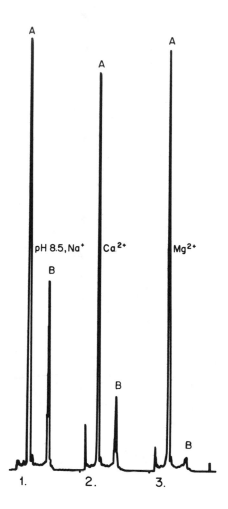

Figure 5. Same as Figure 4 at pH 8.5

QTLC was particularly useful in a study conducted on efrotomycin, an antibiotic produced by Streptomyces lactamdurans (4). Due to the significant difference in the polarity of its isomers (5), only time-consuming and quite tedious gradient HPLC methods offered adequate resolution. For some kinetic studies,

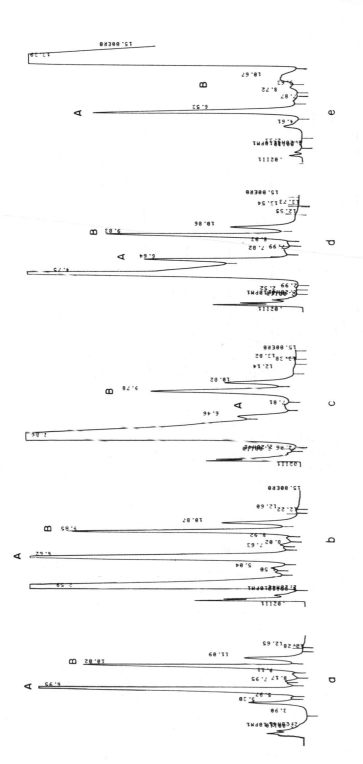

Figure 6. HPLC recording of samples dissolved in different solvents for injection: isopropanol (a), ethyl acetate (b), isopropylacetate (c), methylisobutyl-ketone (d), toluene (e, half the concentration of a–d). Stationary phase: PRP-1 (Hamilton Co.). Mobile phase: linear gradient from 30 to 75 percent solvent B in A during 15 min. A: 0.01 M ammonium phosphate pH 7.5. B: acetonitrile. Wavelength: 230 nm, room temperature. Flow rate: 1.50 ml/min.

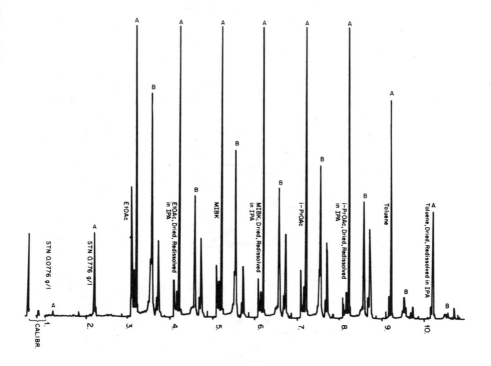

Figure 7. Automatic QTLC scan of the samples shown by Figure 6. The chromatographic and scanning conditions are the same as for Figure 4.

therefore, a much more time-efficient QTLC procedure was developed, increasing the sample through-put of three per hour for HPLC to ten per hour for QTLC. The chromatographic conditions were as follows: adsorbent, silica gel 60F-254 TLC plates (E. Merck); mobile phase, chloroform-methanol-conc. NH_4OH (80:20:1 v/v). All the isomers were detectable at 235 mm (230 nm in HPLC). For some special purposes, when only the A isomer is assayed, 325 nm is a specific wavelength with little or no interference of other isomers present. The principal isomers are called A and B, A being weakly acidic, B, due to an intramolecular ring closure (5), neutral. The kinetics of the conversion of isomer A to isomer B is shown in Figure 3 at two different pH values and at room temperature. The situation is

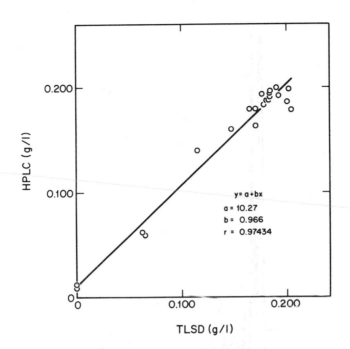

Figure 8. Correlation between the HPLC and QTLC assays of an unidentified bioactive component.

further complicated by the fact that, in addition to pH, certain metal ions also influence the isomerization. Under otherwise identical conditions, at higher pH the A form is more favored than B. The same effect is enhanced by Mg^{2+} and, to a lesser extent, by Ca^{2+}. Na^+ has the least influence among the cations investigated. Figure 4 for pH 7.0 and Figure 5 for pH 8.5 illustrate the effect of the cations mentioned above.

During process development work, efrotomycin had to be assayed in a variety of solvents. It is known that the solvent used for injecting samples during HPLC assay can affect the results. The same sample dissolved in solvents such as isopropanol (Figure 6a), ethyl acetate (Figure 6b), isopropylacetate (Figure 6c) and methylisobutylketone (Figure 6d) resulted in various degrees of peak distortion. While isopropanol caused no interference, some other solvents made the

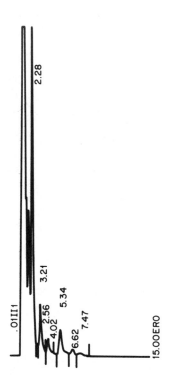

Figure 9. HPLC recording of the unidentified bioactive component (cf. Figure 8), 5.34 min at 230 nm. Stationary phase, μ Bondapack C_{18} (Waters Assoc.); mobile phase, 40 percent (by volume) acetonitrile in 0.01 M ammonium phosphate pH 7.5; flow rate, 1.50 ml/min, room temperature.

chromatograms entirely misleading. Due to limited solubility, the solution in toluene (Figure 6e) contained only half the amount dissolved in the previous solvents. In order to alleviate the problem, the samples had to be evaporated and redissolved in isopropanol for HPLC. QTLC eliminated the requirement for sample preparation. Depending on the polarity of the solvent used for sample application, the zone spreading was somewhat affected, causing the peak shape to change accordingly (Figure 7). However, the impact on the

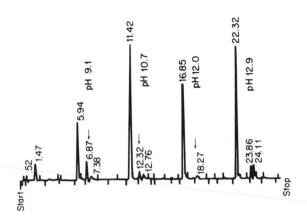

Figure 10. Automatic QTLC scan of five samples of the
unidentified bioactive component (cf. Figure 8) at 235
nm following chromatography in ethyl acetate on silica
gel 60 F-254 TLC plates (E. Merck). Standard: 1.47
min (marked by the arrows). Broth samples 6.87, 12.32
min, trace rejected by the integrator. Instrument:
Shimadzu Model CS-920 combined with Hewlett-Packard
3390A integrator.

resolution of the components and the quantitative
results was not great enough to have any effect on the
conclusion.
        Finally an example is presented in which a UV
absorbing fraction of a fermentation broth is sus-
pected to be carrier of bioactivity. An analytical
procedure was needed to monitor purification steps for
the isolation of enough material for further characterization.
        The correlation between the HPLC (at 230 nm) and
QTLC (at 235 nm) data obtained from assays of whole
broth samples was found satisfactory (Figure 8). The
HPLC recording of a sample heated for 18 hr at 70° C
and pH 10.7 is shown by Figure 9. The fraction of
interest appears at 5.34 min retention time. The same
sample was scanned on the QTLC instrument. The automatic scan of 5 lanes on a 10 x 20 mm plate displays
the following samples from start to stop (Figure 10):
0.1 g/l solution of standard (1.47 min), pH 9.0 sample (6.87) min, pH 10.7 sample (12.32 min), pH 12.0

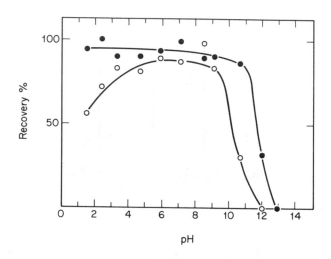

Figure 11. Stability of the unidentified bioactive component as a function of pH at room temperature during
24 hr (●———●), and at 70° C during 18 hr (o———o).

sample (trace, rejected by the integrator), and pH
12.9 sample (none detectable). The purpose of the
experiment was to assess the pH tolerance of the sus-
pected bioactive component at 70° C and at room temperature. The decision was based on the result displayed
by Figure 11 according to which the material was sufficiently stable for processing the broth in the pH
range of 5-9 at 70° C and 1.5-10.5 at room temperature.

CONCLUSIONS

HPLC and QTLC have a great many features in common.
Both allow the quantitative evaluation of chromatograms in a nondestructive manner leaving the sample
intact for further characterization (IR, NMR, MS, bio-
assay, etc.). HPLC is ahead in terms of automation,
although more recently automatic TLC spotting devices
and TLC scanners also appeared on the market. HPLC
started with linear detector systems. Instruments for
QTLC also equipped with the same started appearing in
the early 1970s.

The advantages QTLC can offer over HPLC are as follows:

1. Quantitative evaluation of the chromatograms is possible, even if the UV cutoff of the mobile phase is above the wavelength of the scan.
2. Reduced sensitivity toward the solvent and toward soluble and/or suspended impurities, corrosive chemicals, and so on.
3. Fractions irreversibly bound to the stationary phase can still be detected.
4. Following a scan, the plate can be subjected to further treatment (e.g., spray or gaseous reagents, heating, etc.) followed by yet another quantitative evaluation. For HPLC, postcolumn derivatization would be the corresponding technique requiring expensive additional instrument components, and with greater restrictions in terms of the selection of the reagents.
5. QTLC works on a time-sharing basis. The chromatography can be carried out independently from the instrument, hence many samples can be prepared simultaneously. This means that the rate-limiting factor in sample through-put is the speed of scanning. This particular feature opens up virtually unlimited possibilities for further development of QTLC and its utilization for industrial screening programs. Based on available technology, an instrument could be assembled right now to quantitatively evaluate chromatograms at a speed of one per a few seconds.
6. It is not necessary to dedicate a scanner to one particular assay. In HPLC the conditions of separation, stationary, and mobile phase in particular, are fixed and changes usually take considerable time. In QTLC the adjustment of wavelength and scanning parameters can be accomplished in less than a minute.
7. The presence of few moving parts and mechanical components in the scanner can cause less downtime due to failure.

Advantages of HPLC over QTLC:

1. Higher sensitivity for the same sample volume.
2. Better chromatographic resolution.
3. Advanced automation of the chromatography as
   well as data handling. This is however not an
   inherent advantage. There is no reason to
   assume that the virtually complete automation
   of QTLC can not be accomplished.

    In our laboratories both HPLC and QTLC are
heavily used. The practice that allows the optimum
utilization of available instrumentation is determined
by the fact that the TLC scanner is working on a time-
sharing basis. Whenever its sensitivity and resolution are satisfactory, we use QTLC and liberate instrument capacity for tasks requiring HPLC. The proper
use of HPLC and QTLC side by side allows us to
improve the utilization of instrument time and economy
without compromising on the reliability of our results.

REFERENCES

1. L. R. Treiber, J. Chromatogr. 213, 129 (1981).
2. R. W. Burg, B. M. Miller, E. E. Baker, J. Birnbaum,
   S. A. Currie, R. Hartman, Y.-L. Kong, R. L. Monag-
   han, G. Olson, I. Putter, J. B. Tunac, H. Wallick,
   E. O. Stanley, R. Oiwa, and S. Omura, Antimicrob.
   Agents, Chemother. 15, 361 (1979).
3. T. W. Miller, L. Chaiet, D. J. Cole, L. J. Cole,
   J. E. Flor, R. T. Goegelman, V. P. Gulin, H. Joshua,
   A. J. Kempf, W. R. Krellwitz, R. L. Monaghan, R. E.
   Ormond, K. E. Wilson, G. Albers-Schonberg, and I.
   Putter, Antimicrob. Agents, Chemother. 15, 368
   (1979).
4. R. Wax, W. Maiese, R. Weston, and J. Birnbaum, J.
   Antibiotics 29, 670 (1976).
5. R. S. Dewey, Internal communication, Merck Sharp &
   Dohme Research Laboratories, Rahway, N. J., January
   30, 1976.

CHAPTER 27
Direct Determination of Plasmalogens on Thin Layers

Sidney S. Levin
Kimberly A. Snyder
Joseph C. Touchstone

INTRODUCTION

The group of compounds first found by Feulgen and Voit
in 1924 (1) were called plasmalogens since they
occurred in the plasma of cells. They are glycerophos-
pholipids, i.e., the alk-1-enyl glycero ethers of neutral glycerides. They are widely distributed in
nature in animal and anaerobic bacterial cells.
Almost 20 percent of the phospholipids in the central
nervous system of the human adult are plasmalogens.
The myelin sheath phospholipids are 30 percent plasmalogen. The highest concentration of the plasmalogens
is in the striated muscle of the heart (32 percent).
A 1:4 ratio, plasmalogen to phospholipid, occurs in
the adipose tissue. These proportions vary within
each tissue class and from species to species. The
ethanolamine plasmalogen varies from a low of 23 percent of the phospholipid in the retina of the frog to
75 percent in the sciatic nerve of the rat.
Very little is known about the plasmalogens.
They are found just inside the cell membrane and are
in the highest concentration in the most metabolically
active tissue. Their actual function may be related
to the differences in physical properties between

alkenylacyl and diacylphospholipids or the phospholipid turnover in the membrane.

Methods for the identification and quantitation
of the vinyl-ether linked lipids, particularly in the
plasmalogens, have been reported and widely used.
These methods have been reviewed by Snyder (2). The
procedures described usually depend on the susceptibil-
ity of the enol ether bond to electrophyllic attack.
Analytical procedures for ether lipids not containing
phosphorus are more general because of the difficulty
in separating polar lipids. Treatment of alk-1-enyl-
glycerolipids with hydrogen chloride produces aldehydes (3, 4). Dawson (5) hydrolized the vinyl part of
the plasmalogens with mercuric chloride directly on
the thin layer chromatogram. This was followed by
Schiff's Reagent to detect the aldehydes formed by the
reaction. However, it has been found that the mercuric chloride method resulted in reaction with some
unsaturated derivatives to give several reaction products (6, 7). Another method uses dinitrophenylhydrazine to derivitize the aldehyde released by acid
hydrolysis. The hydrochloric acid and phenylhydrazine
are made up in a single solution (8). Goldfine et al.
have used 90 percent acetic acid to hydrolize plasmen-
ylethanolamine (9).

An improved method for hydrolysis of the vinyl-
ether linkage of plasmalogen is described. It is an
in situ hydrolysis method that attacks only the enol-
ether double bond. Alkyl and unsaturated analogues
are not affected. Treatment of the sample applied to
the thin layer chromatogram with solutions of trichloroacetic acid in dilute hydrochloric acid hydrolyzes
the vinyl-ether linkage. The chromatogram is then
developed and visualized using copper sulfate charring
(1). Differential quantitation using direct densitometry gives the amount of vinyl-ether lipid present in
the sample. The results of the method as applied to
seminal fluid analysis are given.

METHODS

All of the phospholipids used in this study are synthetic. However, the plasmalogens are available only
from natural sources. Phosphatidyl choline and phosphatidyl ethanolamine plasmalogens as well as phospholipids were obtained from Avanti Biochemicals

(Birmingham, Alabama). Ethanolamine plasmalogen was
furnished by Supelco (Bellefonte, Pennsylvania)
through the courtesy of Dr. Lloyd Whiting. A mixture
of the two plasmalogens extracted from beef heart was
obtained from Dr. Howard Goldfine (University of Pennsylvania).

Whatman LK-5 (Clifton, New Jersey) 20 x 20 cm,
250 μ thick silica gel layers with a double thick preabsorbent area are used for all chromatography. All
of the plates are scored into 10-mm lanes with a
Schoeffel scoring device, then washed by developing
overnight in a 1:1 (v/v) $CHCl_3$:MeOH solution. The
washed plates are allowed to dry completely, then used
immediately or stored in a Camag Trockengestell until
used. EM Sciences chromatographic grade glass distilled solvents are used in the preparation of all
mobile phases. Chloroform:methanol (1:1) is used for
predevelopment. The mobile phase used for developing
the chromatogram is chloroform:ethanol:triethylamine:
water (34:30:30:8) (10). The compounds used for
hydrolysis and detection are of reagent grade. They
include trichloroacetic acid (TCA), dinitrophenylhydrazine, cupric sulfate, mercuric chloride and Schiff's
reagent. The cupric sulfate solution is prepared as
a 10 percent w/v solution in 8 percent v/v phosphoric
acid. The TCA is prepared as a 2 percent w/v solution
with 2 percent v/v HCl.

Each sample is applied as a streak across the
lane on the preabsorbent area of the washed plate.
The samples are applied in duplicate on alternate
lanes on each of two plates. One lane of each pair is
treated with 25 μl of TCA/HCl applied dropwise as an
overspot on the still wet sample. This is allowed to
react for 10 min at ambient temperature before drying
completely with forced warm air from a hair dryer set
at low temperature setting. The chromatogram is predeveloped in an uncovered trough or tank, and the solvent front is allowed to move only as far as the preab-
sorbent/sorbent interphase. This procedure extracts
the material and deposits it as a narrow band at the
interphase. The chromatogram is dried completely
after each development. It is then placed in the
mobile phase in an unlined chromatography tank until
the solvent front is within 2 cm of the top of the
plate.

The chromatogram is allowed to dry until the tri-
ethylamine (TEA) cannot be detected. The plates are

first placed on a warm surface to evaporate most of
the TEA, then dried in a 180° oven for 2 min. When
cool, the chromatograms are sprayed with 10 percent
cupric sulfate in 8 percent phosphoric acid (v/v)
until the sorbent is saturated. The copper sulfate is
allowed to react with the sample at ambient temperature for 5 min. The plate is then placed in a 120°
oven for 5 min to remove a major portion of the water.
Charring is carried out in a 170° oven for 10 min.
The chromatograms are cooled, then scanned with a
Kontes model 800 densitometer in double beam, transmission mode. The light source is a white phosphor with
an emission peak of 440 nm and a 300 nm bandwidth.
The densitometer is interphased with a Hewlett Packard
3385A integrator recorder. Quantitation is performed
by the use of calibration curves prepared from synthetic standards of each compound. The concentration
of plasmalogen is calculated by the difference in area
in the plasmalogen zone between the treated and
untreated samples. The hydrolysis reduces or elimi-
nates the area.

Other methods of hydrolysis and detection were
also studied. In one of these the TCA was sprayed
over the developed chromatogram and allowed to dry
before being sprayed with Schiff's reagent to detect
the aldehydes released.

In another procedure mercuric chloride solution
0.05 M (50 µl) is applied to the sample in the preabsorbent region and then Schiff's reagent is sprayed
for detection. This is a modification of a method
described by Owens (6).

The Skipski and Barclay method (8) involves the
simultaneous cleavage and hydrazone formation. A solution of 0.4 percent dinitrophenylhydrazine in 2 N HCl
is sprayed on the developed chromatogram which is then
heated in a 110° oven for 2 min. Yellow zones indicate hydrazone formation. All of these procedures presented problems of reproducibility.

RESULTS AND DISCUSSION

The TCA/HCl reaction acts specifically on the vinyl
C-C linkage; alkyl and unsaturated analogs are not
affected. Table I shows the effect of hydrolyzing
plasmalogen samples obtained from the various sources.
The plasmalogens that are available commercially are

TABLE I. RESULT OF TCA/HCl HYDROLYSIS

Plasmalogen	TCA/HCl, Percent	Source	Assay, Percent
Choline plasmalogen (PCP)	30 plas 70 other	(Avanti)	30 70
Ethanolamine plasma- logen (PEP)	70 plas 30 other	(Avanti)	70 30
Beef heart mix (PCP)	30 plas 70 other	(Goldfine) (Goldfine)	30 70
(PEP)	50 plas 50 other		50 50

natural rather than synthesized preparations, and
therefore are a mixture of plasmalogen and a related
phospholipid, for example, choline plasmalogen is 30
percent plasmalogen, 70 percent phosphatidyl choline.
The table compares the experimental results using this
method with the information furnished by the supplier.
The results for the beef heart extract, which is a
mixture of phosphatidyl choline plasmalogen and phosphatidyl ethanolamine plasmalogen, was furnished by Dr.
Howard Goldfine. Further evidence of the specificity
of the TCA/HCl reaction for the vinyl linkage may be
seen in Table II. The TCA/HCl method is compared with
the $HgCl_2$ method of Owens (6) on unsaturated synthetic
phospholipid standards. The TCA/HCl or the $HgCl_2$ was
applied to the samples as described above. There was
no reaction for any of the samples exposed to the TCA/
HCl. However, when the $HgCl_2$ method was used, three
new products could be seen on the chromatogram of
dioleoyl phosphatidyl choline and complete destruction
of the dilinolenyl phosphatidyl choline. Since the
fatty acid side chain could be any one or a mixture of
these, the $HgCl_2$ would not be useful since it appears
to attack any unsaturated link.
       The acid cleavage with hydrazone formation (8) of
Skipski et al. could not be quantitated with the densitometer since the resulting background was too dense.

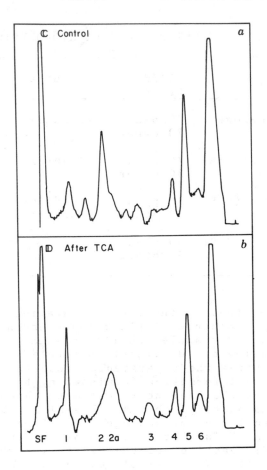

Figure 1. Densitometer scan of a chromatogram sprayed
and charred with CuSO$_4$. (a) Seminal fluid, (b) seminal fluid treated with TCA. SF = solvent front; 1 =
unknown; 2 = phosphatidyl ethanolamine; 3 = lysophosphatidyl ethanolamine; 4 = phosphotidyl choline (lecithin); 5 = sphingomyelin; 6 = lysolecithin.

Figure 1 (a,b) is the scan of a chromatogram from
the analysis of seminal fluid using the present method.
In Figure 1a (top), the diluted seminal fluid is
applied untreated to the plate; in Figure 1b (bottom),

TABLE II

Phospholipid	TCA/HCl	HgCl$_2$
Dioleoyl phosphatidyl choline	No reaction	Three products
Dilinolenyl PC	No reaction	Destroyed
Dipalmitoyl PC	No reaction	No reaction
Dimyristyl PC	No reaction	No reaction

TABLE III. COMPARISON OF HYDROLYSIS METHODS

Method	Percent Hydrolysis
2 percent TCA/8 percent HCl	Total destruction
2 percent TCA	68
5 Percent HCl (v/v)	72
HCl (fumes)	Total destruction

an aliquot of the same sample is applied followed by
TCA/HCl as described. The decrease in the height of
peaks 2 and 4 and the shift of peaks to a new position
(2a and 6) indicate the presence of a number of plasmalogens (vinyl-ether containing substances) in this sample since the peaks in the untreated chromatogram are
not present in the chromatogram obtained after TCA/HCl
treatment.
      The presence of plasmalogens and unsaturated phos-
pholipids in seminal fluid is generally known (11, 12,
13). The methods used for their assay are tedious and
include two-dimensional chromatography as well as
hydrolysis and gas chromatography.
      Some of the modifications of the TCA/HCl and HCl
hydrolysis methods are listed in Table III. These
methods were applied to ethanolamine plasmalogen furnished by Dr. Lloyd Whiting. This is a relatively
pure preparation. The present method resulted in the
90.3 percent hydrolysis, the TCA alone gave only 68
percent, and the 5 percent HCl v/v alone was only 72
percent effective. When the chromatogram was placed

above fuming HCl for 5 min, there was complete destruction of the sample.

The method described provides reproducible
results with a minimum amount of manipulation for the
in situ reaction, separation, and direct densitometric
quantitation of plasmalogens.

The addition of this method to the previously
reported methods (14) for the differentiation of the
saturated and unsaturated fatty acid side chains permits in situ analysis of phospholipids. In this way,
a direct densitometric method is provided for saturated, unsaturated, and the vinyl ether analogs of the
many phospholipids.

REFERENCES

1. H. Debuch and P. Seng, in Ether Lipids: Chemistry
   and Biology, F. Snyder (Ed.), Academic Press, New
   York, 1972, p. 1.
2. F. Snyder, in Lipid Chromatographic Analysis, vol.
   1, 2nd ed., G. V. Marinetti (Ed.), Dekker, New
   York, 1976, p. 111.
3. R. E. Anderson, R. D. Garrett, M. L. Blank, and
   F. Snyder, Lipids 4, 327 (1969).
4. L. A. Horocks, J. Lipid Res. 9, 469 (1968).
5. R. M. C. Dawson, Biochem. J. 75, 45 (1960).
6. K. Owens, Biochem. J. 100, 354 (1966).
7. M. H. Hack and V. J. Flvans, Z. Physiol. Chem. 315,
   157 (1959).
8. V. P. Skipsky and M. Barclay, Methods in Enzymol-
   ogy 14, 530 (1969).
9. H. Goldfine, N. C. Johnston, and M. C. Phillips,
   Biochemistry 20, 2908 (1981).
10. J. C. Touchstone, J. C. Chen, and K. E. Beaver,
    Lipids 15, 61 (1980).
11. A. Darin-Bennet, I. G. White, and D. O. Hoskins,
    J. Reprod. Fert. 49, 119 (1977).
12. R. W. Evans and B. P. Setchell, J. Reprod. Fert.
    57, 189 (1979).
13. D. P. Sclivonchik, P. C. Schmid, V. Natarajan, and
    H. H. O. Schmid, Biochim. Biophys. Acta 618, 242
    (1980).
14. J. C. Touchstone, S. S. Levin, M. F. Dobbins, and
    P. J. Carter, J. High Res. Chromatogr. Chromatogr.
    Comm. 4, 423 (1981).

CHAPTER 28

Separation and Quantitation of Anionic, Cationic, and Nonionic Surfactants by Thin Layer Chromatography

Daniel W. Armstrong
Gale Y. Stine

INTRODUCTION

The analysis of surfactants (e.g., detergents, soaps, etc.) can be a difficult analytical problem. Surfactants are generally somewhat soluble in both water and organic solvents. They concentrate at interfaces and tend to bind to anything available (1, 2). There are a variety of spectrometric, titrimetric, atomic absorption spectrometric, and ion-selective electrode methods for the analysis of surfactants (3-11). All of these techniques have the characteristic of being selective for certain functional groups. For example, both the sodium dodecylsulfate electrode and the methylene blue complex spectrophotometric methods are selective for surfactants with sulfate or sulfonate functional groups. Consequently these techniques give positive responses for a variety of homologous, isomeric, and even structurally dissimilar anionic surfactants. Another shortcoming of these techniques is that one class of surfactants cannot be effectively analyzed in the presence of another. The so-called

neutralization effect of cationic with anionic surfactants is well documented (12). As a result of these
limitations, the analyst has increasingly turned to
physicochemical techniques that provide information on
the total surfactant content in a sample (13) or to
chromatography (14-16). Because most surfactants are
nonvolatile without derivatization, LC or TLC methods
are often preferred. The use of TLC to separate a mixture of anionic surfactants was recently demonstrated
(17). In this work we not only demonstrate the separation of identically charged surfactants from each
other but also the TLC separation of the three major
classes of surfactants (i.e., anionic, nonionic, and
cationic).

   MATERIALS

Whatman reversed-phase TLC plates (KC18F), silica gel
plates (K6F), and hybrid Multi-K plates (CS5) were
activated at 115° C for 2 hr before use. Cetyltrimethylammonium bromide (CTAB, Sigma), cetylpyridinium chloride (CPC, Sigma), cetyltrimethylammonium chloride
(CTAC, Pfaltz & Bauer), dodecylamine (DA, Aldrich),
octadecylamine (OA, Eastman), sodium dodecylsulfate
(SDS, Bio Rad), dodecylbenzenesulfonate (DBS, Pfaltz &
Bauer), sodium dioctylsulfosuccinate (SDOS, Aldrich),
and sodium laurate (SL, Pfaltz & Bauer) were recrystal-
lized three times from ethanol-water before use. The
nonionic surfactants Triton X 100 (TX 100, Bio Rad),
Surfynol 465 (S 465, Air Products), and Igepol CO-530
(ICO-530, GAF) were used as received. ICO-530 is non-
ylphenoxypoly(ethylenoxy)ethanol where the hydrophilic
poly(ethyleneoxy)ethanol "head-group" averages five
units in length. TX 100 is dodecylphenoxypoly(ethyleneoxy)ethanol. S 465 is a poly(ethyleneoxy)ethanol
(averaging 10 units) adduct of 2,4,7,9-tetramethyl-5-
decyn-4,7-diol. Gold label sodium tetraphenylborate
(Aldrich) was used as received. Methanol, ethanol,
methylene chloride, and glacial acetic acid (Baker)
were also used as received.

   METHODS

All separations were done in a 11-3/4-in.-long, 4-in.-
wide, and 10-3/4-in.-high sealed chromaflex developing

tank. The plates were not pre-equilibrated with solvent vapor before use.

Separation of Anionic Surfactants

One ml of 0.1 M SDS, SL, DBS, and SDOS was spotted 1
cm from the bottom of a 5 x 20 cm silica gel plate.
The mobile phase consisted of 8:1 (v/v) methylene
chloride:methanol. The addition of very small amounts
of acetic acid to the mobile phase tended to increase
the R_f's but did not affect the resolution. Spots
were visualized by exposure to I_2 vapor.

Separation of Cationic Surfactants

One µl of 0.1 M CPC, OA, DA, and CTAC or CTAB was spotted 1 cm from the bottom of a 5 x 20 cm silica gel
plate. The mobile phase consisted of 8:1:0.75 (v/v/v)
methylene chloride:methanol:acetic acid. Spots were
visualized by exposure to I_2 vapor.

Separation of Nonionic Surfactants

One µl of 10 percent TX 100, ICO-530 and S 465 were
spotted on a 5 x 20 cm reversed phase (C_{18}) plate.
The mobile phase consisted of 8:2 (v/v) ethanol:2 percent sodium tetraphenylborate$_{(aq)}$. The purpose of
sodium tetraphenylborate was to prevent the spots from
streaking. I_2 vapor was used for visualization.

Separation of Anionic, Cationic, and Nonionic Surfactants

A Whatman CS5, Multi-K, KC18F/K5F 20 x 20 cm plate was
predeveloped in ethanol and then activated at 115° C
for 2 hr. Each surfactant mixture was spotted (0.5 µl)
at a point on the reversed-phase strip. The entire
20 x 20 cm plate was then developed with 75 percent
ethanol in the direction of the reversed-phase strip.
Development was stopped when the solvent front was 2
cm from the top of the plate. Under these conditions,
all anionic surfactants travel at or very near the solvent front (i.e., <2 cm), all cationic surfactants
remain at or near the origin of the reversed-phase
strip (<2.5 cm), while the nonionic surfactants separate between the anionics and cationics. The 20 x 20

cm plate is then cut into three separate sections in a
direction perpendicular to the first development. The
first cut should be 2.5 to 3 cm below the solvent
front. This will isolate the anionic surfactants.
The second cut should be 3 cm above the origin. This
will isolate the cationic surfactants. Perpendicular
secondary development of the plates containing the cationic and anionic surfactants (after reactivation of
the plates) will give complete separation of these species. The mobile phases for secondary development are
8:1 (v/v) $MeCl_2$:MeOH for the anionic surfactants and
8:1:0.5 (v/v/v) $MeCl_2$:MeOH:HOAc for the cationic surfactants. If one develops the entire plate in the
second direction without isolating the anionic and
cationic surfactants as indicated, the nonionic surfactants tend to spread and coat the silica gel portion
of the plate, thereby obscuring all other components.
Visualization is with I_2 vapor.

Quantitation of Surfactants

Scanning densitometry was done with a Shimadzu Model
910 instrument. Surfactants could be detected
directly in the absorbance-reflectance mode at 215 nm.
Detection limits were lower when the developed plate
was exposed to I_2 vapor and scanned at 405 nm (in the
absorbance-transmittance mode).

RESULTS AND DISCUSSION

One's approach to the TLC separation of surfactants in
a mixture is largely controlled by the charge of the
surfactant head-groups as well as the diversity of the
sample. Silica gel is adequate for the separation of
anionic or cationic surfactants from other identically
charged species. Nonionic surfactants are best separated by reversed-phase TLC (RPTLC). Even in RPTLC
nonionic surfactants tend to streak unless a "lipo-
philic salt" such as sodium tetraphenylborate is added.
Table I summarizes the separation conditions for each
class of surfactants. The R_f's of the cationic surfactants can be altered (i.e., increased) considerably
with a slight increase in the concentration of acetic
acid in the mobile phase. The separation of surfactants with identical hydrophylic head-groups (i.e., DA

TABLE I. EXPERIMENTAL CONDITIONS AND R_f VALUES
OF INDIVIDUALLY SEPARATED ANIONIC, CATIONIC,
AND NONIONIC SURFACTANTS

Compound	Stationary Phase	Mobile Phase	R_f
Anionic surfactants	a	c	
1. SDS			0.15
2. DBS			0.09
3. SL			0.70
4. SDOS			0.28
Cationic surfactants	a	d	
1. CTAB			0.21
2. CTAC			0.20
3. CPC			0.27
4. DA			0.42
5. OA			0.55
Nonionic surfactants	b	e	
1. TX 100			0.54
2. S 465			0.70
3. ICO-530			0.45

a Silica gel.
b C_{18} reversed phase.
c 8:1 (v/v) $MeCl_2$:MeOH.
d 8:1:0.75 (v/v/v) $MeCl_2$:MeOH:HOAc.
e 8:2 (v/v) EtOH:2 percent sodium tetraphenylborate$_{(aq)}$.

and OA or Tx 100 and ICO-530) is dependent on the size
of the hydrophobic "tail." Generally the larger the
hydrophobic portion of the surfactant, the greater the
R_f.
    The analysis of solutions containing surfactants
of different charge can be a difficult process because
of precipitation and "neutralization" effects (12).
RPTLC, however, can be used to separate surfactants by
class (see Figure 1). A 75 percent ethanol mobile
phase tends to carry anionic surfactants with the solvent front and leave cationic surfactants near the origin. Perpendicular secondary development of plate

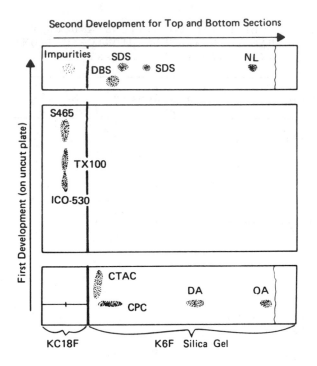

Figure 1. Schematic of a two-dimensional TLC separation of 11 surfactants on a composite reversed-phase
silica gel plate. The first development (on the
reversed-phase strip) separated the surfactants according to class. Secondary development of the top and
bottom sections of the plate results in complete separation of individual surfactants. SDS = sodium
dodecylsulfate, DBS = dodecylbenzenesulfonate, NL =
sodium laurate, S 465 = Surfynol 465, TX 100 = Triton
X 100, ICO-530 = Igepol CO-530, CTAC = cetyltrimethylammonium chloride, CPC = cetylpyridinium chloride, DA =
dodecylamine, OA = octadecylamine.

sections near the solvent front and origin will then
separate the anionic and cationic surfactants into
individual compounds. The secondary development car-
ries the surfactants from the reversed-phase strip
into the silica gel portion of the plate where

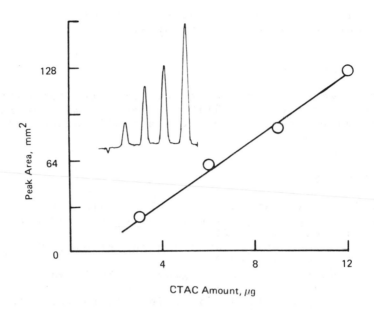

Figure 2. Calibration plot of peak area vs. amount of the standard surfactant (CTAC) chromatographed. The insert shows the actual peaks obtained from scanning densitometry (at 405 nm).

fractionation occurs (Figure 1). Secondary develop-ment of the whole TLC plate or the section of plate containing the nonionic surfactants produced indistin-guishable smears over much of the plate.

Quantitation of surfactants by scanning densitom-etry is a relatively straightforward process. It is possible to directly scan untreated spots at wave-lengths from 200 to 215 nm. Sensitivity and selectiv-ity can be enhanced by using a variety of visualiza-tion or charring techniques (17, 18). Figure 2 shows a scan of four CTAC standards (λ = 405 nm after visual-ization with I_2 vapor) and the corresponding calibra-tion curve.

It is apparent from the literature that exhaus-tive chromatographic separations are presently the most effective means of analyzing complex surfactant mixtures. TLC is shown to be a highly efficient and inexpensive technique for the analysis of a variety of surfactant and surfactant mixtures.

ACKNOWLEDGMENT

This work was supported by grants from the National Science Foundation (CHE-8119055) and Whatman Chemical Separation Division, Inc. We gratefully acknowledge their assistance.

REFERENCES

1. J. H. Fendler and E. J. Fendler, Catalysis in Micellar and Macromolecular Systems, Academic Press, New York, 1975.
2. M. J. Rosen, Surfactants and Interfacial Phenomena, Wiley, New York, 1978.
3. L. K. Wang and D. F. Langley, N. Engl. Water Works Assoc. 89, 301 (1975).
4. L. K. Wang and R. G. Ross, Int. J. Environ. Anal. Chem. 4, 285 (1976).
5. K. Higuchi, Y. Shimoishi, H. Miyata, K. Toei, and T. Yayami, Analyst 105, 768 (1980).
6. L. K. Wang, J. Am. Oil Chem. Soc. 52, 339 (1975).
7. K. Vytras, M. Dajkova, and V. Mach, Anal. Chim. Acta 127, 165 (1981) (and references therein).
8. P. T. Crisp, J. M. Eckert, N. A. Gibson, G. F. Kirkbright, and T. S. West, Anal. Chim. Acta 87, 97 (1976).
9. A. Lebiham and J. Courtot-Coupey, Anal. Lett. 10, 759 (1977).
10. B. J. Kirch and D. E. Clarke, Anal. Chim. Acta 67, 387 (1973).
11. H. M. Rendall, J. Chem. Soc. Faraday Trans. 72, 481 (1976).
12. L. K. Wang and D. F. Langley, N. Engl. Water Works Assoc. 90, 354 (1976).
13. D. W. Armstrong, F. Lafranchise, and D. Young, Anal. Chim. Acta 135, 165 (1982).
14. W. T. Sullivan and R. D. Swisher, Environ. Sci. Technol. 3, 481 (1969).
15. J. K. F. Huber, F. F. M. Kolder, and J. M. Miller, Anal. Chem. 4, 105 (1972).

16. A. Nakae, K. Tsuji, and M. Yamanaka, Anal. Chem. 52, 2275 (1980).
17. C. Yonese, T. Shishido, T. Kaneko, and K. Maruyama, J. Am. Oil Chem. Soc. 59, 2, 112 (1982).
18. G. Zweig and J. Sherman, Handbook of Chromatography, vol. ii, CRC Press, Cleveland, 1972.

Index

A

Accuracy of phospholipid determinations, 297
Accuracy of saliva analysis for nitrite, 220
Acetylated cellulose, 217
Acetylsalicylic acid, 95
Adenosine, 332
Adenosine-5-monophosphate, 131
Advantages, TLC over LC, 77
Agar layers for bioautography, 281
Aminophenols, 103
Analysis of blood for nitrite, 215
Analysis of saliva, blood, 291
Analytical data acquisition system, 170
Angle of scattering of isotopes, 127
Anionic surfactant, 381
Antibiotic analysis, 357

Aromatic hydrocarbons, 180
Asphaltenes, 181
Avermectins, 359

B

Band sharpening, 26
Beer's law, 83
BHT, preservative, 265
Bioautography, 279
Blue tetrazolium, 110
Bonded phases, 25, 87
Breast milk, 262
    Preparation, 262
    Storage, 262
    Neutral lipids, 263
Butylated hydroxytoluene, 263

C

Calibration of densitometer, 80

389